John J. Tobin and Gary Walsh
Medical Product Regulatory Affairs

Related Titles

Walsh, G.

Pharmaceutical Biotechnology

Concepts and Applications

2007
ISBN: 978-0-470-01245-1

Walsh, G.

Biopharmaceuticals

Biochemistry and Biotechnology

2003
ISBN: 978-0-470-84327-7

Bamfield, P.

Research and Development in the Chemical and Pharmaceutical Industry

2006
ISBN: 978-3-527-31775-2

Webster, J. G. (ed.)

Encyclopedia of Medical Devices and Instrumentation

6 Volume Set

2006
ISBN: 978-0-471-26358-6

Zatz, J. L., Teixeira, M. G.

Pharmaceutical Calculations

2004
ISBN: 978-0-471-43353-8

John J. Tobin and Gary Walsh

Medical Product Regulatory Affairs

Pharmaceuticals, Diagnostics, Medical Devices

**WILEY-
BLACKWELL**

WILEY-VCH Verlag GmbH & Co. KGaA

The Authors

Dr. John J. Tobin
ChemHaz Solutions Ltd.
Laccaroe, Feakle
County Clare
Ireland

Prof. Dr. Gary Walsh
Chemical and Environmental Sciences
Department
University of Limerick
Castletroy, Limerick City
Ireland

Library of Congress Card No.: applied for

British Library Cataloguing-in-Publication Data
A catalogue record for this book is available from the British Library.

Bibliographic information published by the Deutsche Nationalbibliothek
Die Deutsche Nationalbibliothek lists this publication in the Deutsche Nationalbibliografie; detailed bibliographic data are available in the Internet at http://dnb.d-nb.de.

Typesetting   Thomson Digital, Noida, India

Cover design   Adam-Design, Weinheim

ISBN: 978-3-527-31877-3

Contents

Medical Product Regulatory Affairs. John J. Tobin and Gary Walsh
Copyright © 2008 WILEY-VCH Verlag GmbH & Co. KGaA, Weinheim
ISBN: 978-3-527-31877-3

Preface

Medical Product Regulatory Affairs aims to introduce and overview the regulatory affairs framework governing the development, approval, manufacturing and surveillance of both pharmaceuticals and medical devices, including *in-vitro* diagnostic products. The book focuses upon the regulatory framework and practice within the European Union and the United States of America, while also outlining global regulatory harmonization measures driven by the International Conference on Harmonization initiative. It should also serve as a reference source for those wishing to work in regulatory affairs, as well as for non-regulatory pharmaceutical/healthcare industry scientists and managers. The scope of this book should also render it a useful reference source for third-level students undertaking healthcare-related programmes of study (e.g. undergraduate or taught postgraduate programs in pharmacy, pharmaceutical science, (bio)materials science, biotechnology or applied biology).

Likewise, it should serve as a useful reference for academic and industry researchers whose research interests relate to the pharmaceutical, diagnostic, or medical device sector.

We would like to gratefully acknowledge EMEA, ISO, EDQM, PIC/S, IMB, ISPE and Health Canada that granted us permission to reproduce selected copyrighted material. We reserve a special acknowledgement and thanks to the European Commission, the FDA and the ICH secretariat, the VICH and the GHTF, who have placed their regulatory documentation in the public domain, and we have reproduced several documents from these sources herein.

Finally we dedicate this book to our parents.

December 2007.

J.J. Tobin
G. Walsh

Medical Product Regulatory Affairs. John J. Tobin and Gary Walsh
Copyright © 2008 WILEY-VCH Verlag GmbH & Co. KGaA, Weinheim
ISBN: 978-3-527-31877-3

1
The Aims and Structure of Regulations

1.1
Introduction

Drugs and medical devices are among the most stringently regulated products in the developed world. This chapter introduces you to the basic principles and concepts behind the regulations so that you can fully appreciate the importance of compliance. The chapter then looks at the general legislative framework that is used to create regulations and identifies the core legal texts that are used to regulate such products in the European Union (EU) and the United States of America (US). Finally, the chapter examines the legal definitions of drugs and medical devices, which are central to determining the scope of the regulations.

1.2
Purpose and Principles of Regulation

The fundamental purpose of regulation is the protection of public health.

Although this appears a very simple goal, its attainment has required the development of extensive and complex regulations. As a newcomer to the subject, you may find some of the regulation cumbersome and overbearing. However, as you study this chapter, you will see that many of the landmark advances in regulatory development were triggered by adverse incidents. Thus, you should accept the current regulations as representing the distilled wisdom of past experience.

To achieve their goal, the regulations rely on a number of core principles and concepts:

Safety
Efficacy
Purpose
Risk/benefit
Quality

Medical Product Regulatory Affairs. John J. Tobin and Gary Walsh
Copyright © 2008 WILEY-VCH Verlag GmbH & Co. KGaA, Weinheim
ISBN: 978-3-527-31877-3

Product safety is an underlying principle for all products. Ideally, the product should do no harm. Thus, the regulations require that the developer or manufacturer must take appropriate steps to demonstrate and ensure the safety of the product under development.

Obviously, for it to be worthwhile, the product must also do some good. Hence, the principle of *efficacy* or effectiveness has become another cornerstone in achieving the goal of regulation. To evaluate effectiveness you must also consider the purpose of the product as expressed in either an *indications for use* statement in the case of drugs, or *intended use* statement in the case of medical devices. As discussed in Section 1.6, and later in Sections 9.3 and 9.4, intended use statements are also vital in determining how some products are regulated in the first place, which in turn dictates the level of scrutiny to which they may be subjected.

In the case of most simple medical devices (a hospital bed for example) it will be relatively straightforward for you to conclude that the product is safe and effective in achieving its intended purpose. However, for more complicated medical devices and many drugs, the situation may not be so clear-cut. Most drugs have some adverse side effects which may range from mild to quite severe. Additionally, many drugs show considerable variation in effectiveness within the patient population that the drug is intended to treat. Thus, you will have to apply the concept of *Risk to Benefit* when making a judgement as to whether a product should be marketed and as to what limitations, if any, should apply to its use. Looking at it from a regulatory stance you must ask the questions, do the benefits outweigh the risks, and in the overall balance does the product enhance public health?

Consideration of the following examples of existing drug products may help you to understand this point. Chemotherapy drugs used to fight cancer are known to have significant side effects, including serve nausea and hair loss, while they are rarely effective in all cancer patients. However, despite their limitations they still provide a vital element in the fight against cancer as they can contribute to the cure of what could otherwise be a fatal disease.

In recent years concerns have been raised in the popular press about possible side effects from the MMR vaccine, which is given to infants to guard against measles, mumps and rubella. Although this has led to a drop in the levels of vaccination, the advice from health professionals continues to be in favour of vaccination, because even if the claimed side effects were shown to be true, failure to vaccinate would still statistically pose the greater health risk due to the detrimental effects of the diseases themselves.

The final element which regulations address is *quality*. Safety and fitness for purpose, as discussed above, are two of the characteristics that you would associate with a quality product. However, these characteristics alone would not describe a quality product. For any product or service to be considered quality you would also expect it to be reliable and consistent. Additionally, in the context of medical products, quality means a requirement to demonstrate conformance to agreed specifications or applicable standards for content, purity and stability. Many organisations, from manufacturers to service providers, voluntarily apply quality assurance systems in order to more effectively meet their customers' needs on a consistent basis. However,

Figure 1.1 Regulatory principles.

this is not a voluntary option for manufacturers of drugs and potential high-risk medical devices. Such enterprises are legally required to apply an appropriate quality assurance system, the specifics of which are, for the most part, defined in regulations. These basic principles are illustrated in Figure 1.1.

1.3
The Legal Framework for Regulation

As you will encounter many different types of legal instruments during the course of this book, it is worthwhile that you take some time to understand the basic principles on which such instruments are constructed.

1.3.1
National Legislative Process

In a modern constitutional democracy, laws are created via a hierarchical legislative process. You will find the principal legal principles laid down in a *constitution*, which derives its legitimacy directly from the will of the people and can only be amended via referendum. The constitution sets out your basic rights as an individual in the state, and establishes a system of governance that provides for legislative, executive and judicial branches of government.

The legislature consists of elected representatives who act on behalf of the people in a legislative assembly (houses of parliament) and have the power to propose new legislation in the form of a *Bill*. In practice, most legislation is introduced by Government Ministers in their role as the political heads of the executive branch of government. After a number of stages during which it is scrutinised and debated, the Bill, if acceptable, is approved by majority vote in the houses of parliament. It then proceeds to become an *Act* once it is signed into law by the head of state.

An Act establishes the broad legal requirements pertaining to a particular topic and grants powers of enforcement to the relevant Government Minister. An Act will also usually confer power on the Minister to issue further detailed regulations that enable practical application and enforcement of the Act. Such regulations are issued in the form of Statutory Instruments in Europe or additions to the Code of Federal Regulations (CFR) in the US.

In summary, you will find that Acts contain the broad legal principles whereas you are more likely to find the detailed technical requirements of the law in the regulations.

The executive branch of government is responsible for executing the law. It consists of the ministerial heads of each government department together with the civil service and all other state agencies and authorities empowered to administer and enforce the law. The judicial branch function as independent guardians of your rights and adjudicate on whether the executive have, in applying the law, overstepped the powers granted to them via the constitution, acts or regulations.

1.3.2
EU Legislative Process

A different system applies to the creation of legislation at EU level. The EU is based on a series of treaties between member states, which are comparable to constitutional law at national level. Three institutions are involved in the creation of EU law: (i) The European Commission; (ii) The Council of the European Union; and (iii) The European Parliament.

The European Commission acts as the executive body and is headed by Commissioners nominated by the member states. It is primarily responsible for preparing and presenting legislative proposals. Responsibility for approval of the proposals is shared between the Council, which consists of the Government Ministers from each member state, the European Parliament, which contains directly elected representatives and the Commission. Different mechanisms for the distribution of power between the institutions are used, depending on the subject matter of the legislation. Approval of basic legislative measures requires the involvement of the Council and the Parliament, whereas the Commission are empowered to approve provisions of a technical or administrative nature. The issuing authority will always be identified in the title of the document.

Binding EU legislation is issued in the form of Regulations, Directives and Decisions.

An *EU Regulation* is directly applicable in each member state, without the need for transposition into national legislation. However, you will find that some supplementary national legislation is usually required so as to establish penalties and powers of enforcement at national level.

Directives, on the other hand, are addressed to member states and require that they enact national legislation so as to achieve the objectives of the directives. Thus, a directive allows flexibility in how national legislation is enacted. In practice, national legislation will frequently refer you back to the directive, particularly when a directive contains large amounts of detailed technical requirements.

Regulations and Directives use a similar structure.

- You will start by reading statements citing the legal basis for the document and the reasoning behind its creation ("whereas" statements).
- Then, you will find the fundamental legal requirements set out in a series of articles.
- Finally, where applicable, you will find detailed technical requirements in one or more Annexes.

In a sense, the articles equate to what you might expect to find in an Act at national level, while the content of Annexes would be more akin to what would be placed in regulations. There is also a parallel in terms of authorisation, in that amendments to the articles usually require the approval of the political institutions, whereas adaptation of the Annexes to technical progress is possible via a decision of the Commission, functioning as the executive body. You can see this in practice by just looking at the title of each instrument that you read.

The final legal instrument is a *Decision*. A decision focuses on an individual measure and is directly binding in its entirety on the specific individuals or entities to whom it is addressed. The Commission uses Decisions to issue marketing authorisations for approval of new drugs granted under a "centralised" procedure (see Chapter 6). Figure 1.2 summarises the relationship between various legal instruments used in Europe.

1.3.3
Working with Legal Texts

It is advisable that, for the most part, you use the EU documents as your primary source of legislation. There are a number of benefits to doing this:

- You get both the principal legal requirements (The Articles) and the technical detail (The Annexes) in one document. As mentioned above, national legislation may just transpose the Articles, and you may have to refer back to the directive for the technical Annexes.

- National legislation is moulded by Directives, and new national legislation is invariably a response to EU initiatives.

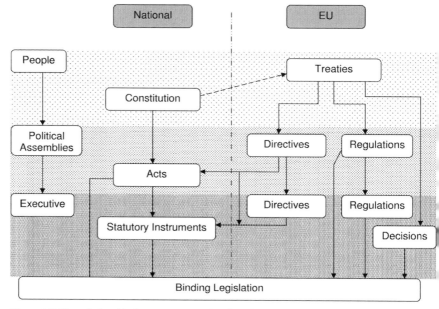

Figure 1.2 The relationship between National and EU legal instruments, and the flow of legislative authority.

- Most products are targeted at international rather than just national markets. Once you comply with the requirements of the directive, national legislation may not impose additional requirements other than as provided for in the directive (language requirements, etc.).

However, when working with Directives, you need to be careful about updates. Once a "base" Directive is established, subsequent Directives can be issued to amend one or more of the Articles of the "base" Directive, or to adapt the Annexes to technical progress. This makes the original section of the base directive no longer applicable. To help you work with the legislation, the EU prepares consolidated texts. However, it is only the Directives as published in the Official Journal of the European Community that have legal standing. Occasionally, in the interests of clarity, the EU will start afresh and recast a new "base" Directive incorporating all previous amendments.

1.3.4
Guidance Documents

In addition to the legal texts, you will also encounter guidance documents issued by the agencies involved in application and enforcement of legislation and other interested parties.

These are intended to help you understand what the law requires and to provide you with solutions as to what to do to meet the requirements. There is considerable variety in the type of guidance documents available. Some documents are used to describe specific requirements in precise detail, such as the procedures for making regulatory submissions, whereas other documents will tend to be more general in nature and may just raise points to consider or suggested approaches. In practice, they are of great practical value and give a very good insight into what an agency is expecting in terms of application of regulations. Guidance documents, adopted pursuant to specific requirements contained in EU Regulations or Directives, have a derived legal status. However, other guidance does not have formal legal status and may not be taken as an interpretation of what the law requires, as such a determination is the preserve of the judiciary. Irrespective of it status, industry are advised to follow all relevant guidance, so as to facilitate smoother interaction with the regulatory authorities, and avoid having to justify alternative approaches that may otherwise be used.

1.3.5
Pharmacopoeia

Pharmacopoeial publications provide a final important source of information for the pharmaceutical industry, regulatory authorities, and the healthcare professions. These are concerned with establishing quality standards. These publications include monographs that define specifications for the purity and identity of established pharmaceutical ingredients, both active and non-active, together with recognised analytical methods that may be used to evaluate them. The most relevant are the *United States Pharmacopoeia* (USP) and the *European Pharmacopoeia* (Ph.Eur).

1.4
Basic Legislation

1.4.1
EU Legislation

The core legislation governing the regulation of drugs in the EU is contained in two "base" Directives, which provide the framework for regulation of medicines at national level. These are:

- 2001/82/EC: Directive 2001/82/EC of the European Parliament and of the Council of 6 November 2001 on the Community code relating to veterinary medicinal products.

- 2001/83/EC: Directive 2001/83/EC of the European Parliament and of the Council of 6 November 2001 on the Community code relating to medicinal products for human use.

The Human Medicines Directive replaced an original directive and its amendments that dated back to 1965 (65/65/EEC). This original directive was prompted by a

Table 1.1 Comparison of the content headings of the Human and Veterinary Medicines Directives.

Veterinary Medicines Directive 2001/82/EC (96 Articles)	Human Medicines Directive 2001/83/EC (130 Articles)
Title I: Definitions	Title I: Definitions
Title II: Scope	Title II: Scope
Title III: Marketing	Title III: Placing on the market
Chapter 1: Marketing authorisation	Chapter 1: Marketing authorisation
Chapter 2: Particular provisions applicable to homeopathic veterinary medicinal products	Chapter 2: Special provisions applicable to homeopathic medicinal products
Chapter 3: Procedure for marketing authorization	Chapter 3: Procedures relevant to the marketing authorization
Chapter 4: Mutual recognition procedure and decentralised procedure	Chapter 4: Mutual recognition procedure and decentralised procedure
Title IV: Manufacture and imports	Title IV: Manufacture and importation
Title V: Labelling and package insert	Title V: Labelling and package leaflet
	Title VI: Classification of medicinal products
Title VI: Possession, distribution and dispensing of veterinary medicinal products	Title VII: Wholesale distribution of medicinal products
	Title VIII: Advertising
Title VII: Pharmacovigilance	Title IX: Pharmacovigilance
	Title X: Special provisions on medicinal products derived from human blood and plasma
Title VIII: Supervision and sanctions	Title XI: Supervision and sanctions
Title IX: Standing committee	Title XII: Standing committee
Title X: General provisions	Title XIII: General provisions
Title XI: Final provisions	Title XIV: Final provisions
Annex I:	Annex I:

determination to prevent a recurrence of a catastrophe that came to light in the early 1960s, when it was concluded that the birth of thousands of babies with limb deformities was as a result of their mothers having taken a new sedative drug, thalidomide, during pregnancy. This proved to be a cathartic event as it exposed the limitations in the regulatory measures that existed at the time, and prompted new legislative measures in many jurisdictions worldwide. The main purpose of the directive introduced in 1965 was to set standards for drug authorisation that should be applied across all member states. The Veterinary Medicines Directive replaced a similar set of directives dating back to 1981. Both directives are similar in structure, with articles grouped under various titles, as shown in Table 1.1. The directives also contain large Annexes that set out the detailed requirements pertaining to the approval of drugs in the EU. A number of amending directives and regulations have already been issued that update the articles and annexes for technical progress (see Table 1.2).

Table 1.2 Updates of the Medicines Directives.

Veterinary Medicines Directive 2001/82/EC Updates	
Dir. 2004/28/EC	Amended by Directive 2004/28/EC of the European Parliament and of the Council of 31 March 2004 amending Directive 2001/82/EC on the Community code relating to veterinary medicinal products
Human Medicines Directive 2001/83/EC Updates	
Dir. 2002/98/EC Human blood products	Amended by Directive 2002/98/EC of the European Parliament and of the Council of 27 January 2003 setting standards of quality and safety for the collection, testing, processing, storage and distribution of human blood and blood components and amending Directive 2001/83/EC
Dir. 2003/63/EC (Annex I update)	Amended by Commission Directive 2003/63/EC of 25 June 2003 amending Directive 2001/83/EC of the European Parliament and of the Council on the Community code relating to medicinal products for human use
Dir. 2004/24/EC Herbal medicines	Amended by Directive 2004/24/EC of the European Parliament and of the Council of 31 March 2004 amending, as regards traditional herbal medicinal products, Directive 2001/83/EC on the Community code relating to medicinal products for human use
Dir. 2004/27/EC	Amended by Directive 2004/27/EC of the European Parliament and of the Council of 31 March 2004 amending Directive 2001/83/EC on the Community code relating to medicinal products for human use
Reg. EC/1901/2006 (Paediatric use)	Regulation (EC) No. 1901/2006 of the European Parliament and of the Council of 12 December 2006 on medicinal products for paediatric use and amending Regulation (EEC) No. 1768/92, Directive 2001/20/EC, Directive 2001/83/EC and Regulation (EC) No. 726/2004
Reg. EC/1902/2006 (Paediatric use)	Regulation (EC) No. 1902/2006 of the European Parliament and of the Council of 20 December 2006 amending Regulation 1901/2006 on medicinal products for paediatric use
Reg. EC/1394/2007 (Advanced therapy)	Regulation (EC) No. 1394/2007 of the European Parliament and of the Council of 13 November 2007 on advanced therapy medicinal products and amending Directive 2001/83/EC and Regulation (EC) No. 726/2004

Some categories of medicinal products require direct regulation from EU institutions. Regulation (EC) No. 726/2004 lays down Community procedures for the authorisation, supervision and pharmacovigilance of medicinal products for human and veterinary use, and establishes a European Medicines Agency. This regulation replaces a previous regulation from 1993 (Regulation No. 2309/93) that initiated this

Table 1.3 Content headings of Regulation (EC) 726/2004.

Title	Topic
I	Definitions & Scope
II	Authorisation and supervision of medicinal products for human use
Chapter 1	Submission and examination of applications — Authorisations
Chapter 2	Supervision and penalties
Chapter 3	Pharmacovigilance
III	Authorisation and supervision of veterinary medicinal products
Chapter 1	Submission and examination of applications — Authorisations
Chapter 2	Supervision and penalties
Chapter 3	Pharmacovigilance
IV	The European Medicines Agency – responsibilities and administrative structures
Chapter 1	Tasks of the agency
Chapter 2	Financial Provisions
Chapter 3	General Provisions governing the Agency
V	General and final provisions

process. A summary of the main topics contained in the regulation is shown in Table 1.3.

Community-wide regulation of medical devices commenced with the introduction of Council Directive 90/385/EEC of 20 June 1990 on the approximation of the laws of the Member States relating to active implantable medical devices. Two further "base" directives followed that cover all other medical devices: The Medical Devices Directive 93/42/EEC and The In Vitro Diagnostics Directive 98/79/EC. All three "base" directives are similar in content and structure. However, it should be noted that, in addition to dealing with the particular subject matter, the Medical Devices Directive and the In Vitro Diagnostics Directive also contained amendments to the previous device directives. The Medical Devices Directive amended articles in the Active Implantable Medical Devices Directive, while the In Vitro Diagnostics Directive amended articles in the Medical Devices Directive.

There have been a number of amending directives since the base directives were issued; these are summarised in Table 1.4. Directive 2007/47/EC is the most important as it contains significant amendments to all three base directives. It builds on the practical experience gained in implementing the directives, and sets out to simplify and harmonise the language of the directives so as to ensure consistent interpretation and application of the requirements in all Member States. Among other items addressed,

- it amends the definition of a medical device so that software can be regarded as a medical device in its own right;
- it enhances the requirements for clinical investigations in line with international developments;
- it updates some of the classification rules for medical devices to achieve greater clarity;

Table 1.4 Updates of the Device Directives.

Directive	Scope
2000/70/EC	Amends Council Directive 93/42/EEC as regards medical devices incorporating stable derivates of human blood or human plasma
2001/104/EC	Contains further clarification on the regulation of human blood or plasma products
2003/12/EC	Reclassifies breast implants as Class III devices by way of derogation from the general classification rules
2003/32/EC	Introduces detailed specifications as regards the requirements laid down in Council Directive 93/42/EEC with respect to medical devices manufactured utilizing tissues of animal origin
2005/50/EC	Reclassifies hip, knee and shoulder joint replacements as Class III devices by way of derogation from the general classification rules
2007/47/EC	Contains a general update and overhaul of all three base directives

- it recognises the advances in information technology that facilitate the distribution of instructions for use by electronic means; and
- and it clarifies that the post-market vigilance reporting system should apply to custom made devices.

There are a number of other regulations/directives that you will need to consult, as appropriate. These address topics such as Good Laboratory Practice (GLP), Good Manufacturing Practice (GMP), the conduct of clinical trials, variations to authorised drugs, and the use of genetically modified organisms. A list of the most relevant directives is shown in Table 1.5.

1.4.2
US Legislation

Regulatory authority in the US derives primarily from the Federal Food, Drug, and Cosmetic Act (FDC Act). The act was originally passed into law in 1938, replacing a previous Food and Drugs Act that dated back to 1906. Impetus for approval of the FDC Act came from the drug-related death of 107 people. The victims, mainly children, had taken a sulphanilamide drug preparation that contained poisonous diethylene glycol as a solvent in order that it could be presented in a more palatable, raspberry-flavoured liquid form. The Act required for the first time that manufacturers test new drugs for safety and submit their results to the Food and Drugs Administration (FDA) for marketing approval. In addition, it authorised the FDA to conduct unannounced inspections of manufacturing facilities. Many amendments to the act have been introduced since then, the single most significant being the Kefauver–Harris amendment of 1962, which introduced the requirement that drugs must be shown to be effective as well as safe. This was the main US response to the thalidomide disaster. An outline of the content of the Act is shown in Figure 1.3. Because of their historical evolution, biologic products are regulated under different

Table 1.5 Selected other directives and regulations of relevance.

2001/20/EC (Clinical practice)	Directive 2001/20/EC of the European Parliament and of the Council of 4 April 2001 on the approximation of the laws, regulations and administrative provisions of the Member States relating to the implementation of good clinical practice in the conduct of clinical trials on medicinal products for human use.
2005/28/EC (Clinical practice)	Commission Directive 2005/28/EC of 8 April 2005 laying down principles and detailed guidelines for good clinical practice as regards investigational medicinal products for human use, as well as the requirements for authorisation of the manufacturing or importation of such products
2003/94/EC (GMP Human)	Commission Directive 2003/94/EC of 8 October 2003 laying down the principles and guidelines of good manufacturing practice in respect of medicinal products for human use and investigational medicinal products for human use.
91/412/EEC (GMP Veterinary)	Commission Directive 91/412/EEC of 23 July 1991 laying down the principles and guidelines of good manufacturing practice for veterinary medicinal products
2004/10/EC (GLP)	Directive 2004/10/EC of the European Parliament and of the Council of 11 February 2004 on the harmonisation of laws, regulations and administrative provisions relating to the application of the principles of good laboratory practice and the verification of their applications for tests on chemical substances
EC/1084/2003 (Variations)	Commission Regulation (EC) No. 1084/2003 of 3 June 2003 concerning the examination of variations to the terms of a marketing authorisation for medicinal products for human use and veterinary medicinal products granted by a competent authority of a Member State
EC/1085/2003 (Variations)	Commission Regulation (EC) No. 1085/2003 of 3 June 2003 concerning the examination of variations to the terms of a marketing authorisation for medicinal products for human use and veterinary medicinal products falling within the scope of Council Regulation (EEC) No. 2309/93
2001/18/EC (GMO release)	Directive 2001/18/EC of the European Parliament and of the Council of 12 March 2001 on the deliberate release into the environment of genetically modified organisms and repealing Council Directive 90/220/EEC
98/81/EC (GMO containment)	Council Directive 98/81/EC of 26 October 1998 amending Directive 90/219/EEC on the contained use of genetically modified micro-organisms
EC/141/2000 (Orphan drug)	Regulation (EC) No. 141/2000 of the European Parliament and of the Council of 16 December 1999 on orphan medicinal products
EC/847/2000 (Orphan drug)	Commission Regulation (EC) No. 847/2000 of 27 April 2000 laying down the provisions for implementation of the criteria for designation of a medicinal product as an orphan medicinal product and definitions of the concepts 'similar medicinal product' and 'clinical superiority'
EEC/2377/90 (MRLs)	Council Regulation (EEC) No. 2377/90 of 26 June 1990 laying down a Community procedure for the establishment of maximum residue limits for veterinary medicinal products in foodstuffs of animal origin.
EC/1308/1999 (MRLs)	Council Regulation (EC) No. 1308/1999 of 15 June 1999 amending Regulation (EC) No. 2377/90 laying down a Community procedure for the establishment of maximum residue limits of veterinary medicinal products in foodstuffs of animal origin
EEC/1768/92 (Patent protection)	Council Regulation (EEC) No. 1768/92 of 18 June 1992 concerning the creation of a supplementary protection certificate for

Chapter I – Short Title
Chapter II – Definitions
Chapter III – Prohibited Acts and Penalties
Chapter IV—Food
Chapter V – Drugs and Devices:
   Subchapter A – Drugs and Devices:
   SEC. 501.   ADULTERATED DRUGS AND DEVICES
   SEC. 502.   MISBRANDED DRUGS AND DEVICES
   SEC. 503.   EXEMPTIONS AND CONSIDERATION FOR CERTAIN DRUGS, DEVICES, AND
                BIOLOGICAL PRODUCTS
   SEC. 503A.   PHARMACY COMPOUNDING.
   SEC. 504.   VETERINARY FEED DIRECTIVE DRUGS
   SEC. 505.   NEW DRUGS
   SEC. 505A.   PEDIATRIC STUDIES OF DRUGS
   SEC. 505B.   RESEARCH INTO PEDIATRIC USES FOR DRUGS AND BIOLOGICAL PRODUCTS.
   SEC. 506.   FAST TRACK PRODUCTS.
   SEC. 506A.   MANUFACTURING CHANGES.
   SEC. 506B.   REPORTS OF POSTMARKETING STUDIES.
   SEC. 506C.   DISCONTINUANCE OF A LIFE SAVING PRODUCT.
   SEC. 508.   AUTHORITY TO DESIGNATE OFFICIAL NAMES
   SEC. 509   NONAPPLICABILITY TO COSMETICS
   SEC. 510.   REGISTRATION OF PRODUCERS OF DRUGS AND DEVICES
   SEC. 512.   NEW ANIMAL DRUGS
   SEC. 513.   CLASSIFICATION OF DEVICES INTENDED FOR HUMAN USE
   SEC. 514.   PERFORMANCE STANDARDS
   SEC. 515.   PREMARKET APPROVAL
   SEC. 516.   BANNED DEVICES
   SEC. 517.   JUDICIAL REVIEW
   SEC. 518.   NOTIFICATION AND OTHER REMEDIES
   SEC. 519.   RECORDS AND REPORTS ON DEVICES
   SEC. 520.   GENERAL PROVISIONS RESPECTING CONTROL OF DEVICES INTENDED FOR
                HUMAN USE
   SEC. 521   STATE AND LOCAL REQUIREMENTS RESPECTING DEVICES
   SEC. 522.   POSTMARKET SURVEILLANCE
   SEC. 523.   ACCREDITED PERSONS.
   Subchapter B – Drugs for Rare Diseases and Conditions
   SEC. 525   RECOMMENDATIONS FOR INVESTIGATIONS OF DRUGS FOR RARE DISEASES
                OR CONDITIONS
   SEC. 526   DESIGNATION OF DRUGS FOR RARE DISEASES OR CONDITIONS
   SEC. 527   PROTECTION FOR DRUGS FOR RARE DISEASES OR CONDITIONS
   SEC. 528   OPEN PROTOCOLS FOR INVESTIGATIONS OF DRUGS FOR RARE DISEASES
                OR CONDITIONS
   Subchapter C – Electronic Product Radiation Control
   Subchapter D – Dissemination of Treatment Information
   Subchapter E – General Provisions Relating to Drugs and Devices
   SEC. 561.   EXPANDED ACCESS TO UNAPPROVED THERAPIES AND DIAGNOSTICS.
   SEC. 562.   DISPUTE RESOLUTION.
   SEC. 563.   CLASSIFICATION OF PRODUCTS.
   SEC. 564.   AUTHORIZATION FOR MEDICAL PRODUCTS FOR USE IN EMERGENCIES.
   Subchapter F—New Animal Drugs for Minor Use and Minor Species
   SEC. 571.   CONDITIONAL APPROVAL OF NEW ANIMAL DRUGS FOR MINOR USE AND
                MINOR SPECIES.
   SEC. 572.   INDEX OF LEGALLY MARKETED UNAPPROVED NEW ANIMAL DRUGS FOR
                MINOR SPECIES.
   SEC. 573.   DESIGNATED NEW ANIMAL DRUGS FOR MINOR USE OR MINOR SPECIES.
Chapter VI – Cosmetics
Chapter VII – General Authority:
   Subchapter A – General Administrative Provisions
   Subchapter B – Colors
   Subchapter C – Fees
   Subchapter D – Information and Education
   Subchapter E – Environmental Impact Review
   Subchapter F – National Uniformity for Nonprescription Drugs and
                   Preemption for Labeling or Packaging of Cosmetics
   Subchapter G – Safety Reports

Figure 1.3 Content of the Food, Drug and Cosmetic (FDC) Act.

Chapter VIII – Imports and Exports

Chapter IX—Miscellaneous

Note: Chapters/sub-chapters of most relevance are highlighted in bold.

Figure 1.3 (*Continued*)

acts, section 351 of the Public Health Services (PHS) Act in the case of biologics for human use and section 151–159 of the Virus-Serum-Toxin Act in the case of veterinary biologics.

Detailed regulations supporting the Acts are published principally in Title 21 of the Code of Federal Regulations (21 CFR). An outline of the main sections of the Title is shown in Table 1.6. Regulations in support of veterinary biologics are contained in Title 9 of the Code of Federal Regulations, Parts 101–123 (see Table 1.7).

Table 1.6 Content of Title 21 of the Code of Federal Regulations.

Volume No	Contents
1	Parts 1 to 99. General regulations for the enforcement of the Federal Food, Drug, and Cosmetic Act and the Fair Packaging and Labeling Act. Color additives.
	Part 11 Electronic Records; Electronic Signatures
	Part 50 Protection of Human Subjects
	Part 54 Financial Disclosure by Clinical Investigators
	Part 56 Institutional Review Boards
	Part 58 Good Laboratory Practice
2	Parts 100 to 169. Food standards, good manufacturing practice for foods, low-acid canned foods, acidified foods, and food labeling.
3	Parts 170 to 199. Food additives.
4	Parts 200 to 299. General regulations for drugs.
	Part 201 Labelling
	Part 207 Registration of Drug Producers & Drug Listings
	Part 210 cGMP Manufacturing, Processing Packing, Holding
	Part 211 cGMP Finished Pharmaceuticals
	Part 225 cGMP Medicated Feeds
	Part 226 cGMP Medicated Articles
5	Parts 300 to 499. Drugs for human use.
	Part 312 Investigational New Drug (IND)
	Part 314 New Drug Marketing Approval Applications (NDA)
	Part 320 Bioavailability and Bioequivalence Requirements
6	Parts 500 to 599. Animal drugs, feeds, and related products.
	Part 511 New Animal Drugs for Investigational Use
	Part 514 New Animal Drug Applications (NADA)
7	Parts 600 to 799. Biologics and cosmetics.
	Part 600 Biologic Products General
	Part 601 Biologic Licence Applications (BLA)
	Part 606 cGMP Blood & Blood Products
	Part 607 Establishment Registration & Product Listing

(*Continued*)

Table 1.6 (*Continued*)

Volume No	Contents
8	Parts 800 to 1299. Medical devices and radiological health. Regulations under the Federal Import Milk Act, the Federal Tea Importation Act, the Federal Caustic Poison Act, and for control of communicable diseases and interstate conveyance sanitation.
	Part 801 Labelling
	Part 803 Medical Device Reporting
	Part 806 Corrections & Removals
	Part 807 Establishment Registration & Device Listing
	Part 809 In vitro Diagnostics (IVDs)
	Part 812 Investigational Device Exemptions (IDEs)
	Part 814 Pre Market Approval (PMA)
	Part 820 Quality System Regulation (QSR)
	Part 822 Market Surveillance
	Part 860 Medical Device Classification Procedures
9	Parts 1300 through end. Drug Enforcement Administration regulations and requirements.

Note: Parts of most relevance for the current book are individually listed

Table 1.7 Contents of Title 9 of the Code of Federal Register dealing with veterinary biologics.

Part	Description
101	Definitions
102	Licenses for biological products
103	Experimental production, distribution, and evaluation of biological products prior to licensing
104	Permits for biological products
105	Suspension, revocation, or termination of biological licenses or permits
106	Exemption for biological products used in department programs or under department control or supervision
107	Exemptions from preparation pursuant to an unsuspended and unrevoked license
108	Facility requirements for licensed establishments
109	Sterilization and pasteurization at licensed establishments
112	Packaging and labeling
113	Standard requirements
114	Production requirements for biological products
115	Inspections
116	Records and reports
117	Animals at licensed establishment
118	Detention; seizure and condemnation
121	Possession, use, and transfer of biological agents and toxins
122	Organisms and vectors
123	Rules of practice governing proceedings under the Virus-Serum-Toxin Act

1.5
Scope of the Legislation

The spectrum of drugs and medical devices covered by the legislation is quite diverse. While many products are easily identified as being subject to the regulations, careful application of the legal definitions of drugs and devices is required to establish the status of other "borderline" products. The definitions of drugs and devices taken from the relevant EU and US legislation are shown in Figure 1.4. (Note: Drugs are referred to as medicinal products in EU legislation.)

In order to determine the regulatory status of an individual product you need to answer two key questions:

- What is it supposed to do?
- How does it do it?

To answer the first question you need to examine the intended use statement for the product and see if it claims a medical purpose corresponding to any of those contained in the definitions.

The key action verbs to look out for are to *treat, prevent, diagnose, cure, mitigate, restore, correct, modify, replace* or *alleviate* a disease or condition.

Once you have established a medical purpose, a careful examination of its primary mode of action will allow you to decide whether the product is a drug or device.

To understand the process more clearly we shall look at the following examples, which illustrate some of the distinctions:

- *Traditional herbal and homeopathic remedies* that are supplied as natural treatments for medical conditions or diseases are subject to regulation as drugs, for example St. John's wort.

- *Health foods* and other *functional foods* that may have beneficial health effects are generally not considered drugs, as their primary purpose is nutritional. However, any information on health benefits must not include specific medical claims that are associated with drug products. Recent European legislation (Regulation (EC) No. 1924/2006 on nutrition and health claims made on foods) sets out to define the types of claims that can be made for such products. Examples of such foods include plant-sterol-containing foods that help reduce cholesterol levels, gluten-free foods that prevent the symptoms of coeliac disease, and pro-biotic yoghurts that promote healthy gut flora.

- *Dietary supplements* supplied in dosage form can present a grey area. In the US, there is specific legislation dealing with dietary supplements including vitamins, minerals, and enzymes. These are excluded from drug regulations, provided that specific drug claims are avoided. In the EU, specific legislation dealing with dietary supplements covers only vitamins and minerals. Enzyme supplements such as lactase, which is used as a digestive aid to treat lactose intolerance, could be viewed as a drug. However, with the advent of authorised health and nutritional claims for functional foods it is more likely to be viewed as a food.

Drugs

A medicinal product for human use is defined in the EU as:

(a) Any substance or combination of substances presented as having properties for treating or preventing disease in human beings;
or
(b) Any substance or combination of substances which may be used in or administered to human beings either with a view to restoring, correcting or modifying physiological functions by exerting a pharmacological, immunological or metabolic action, or to making a medical diagnosis.

A veterinary medicinal product is similarly defined as:

(a) Any substance or combination of substances presented as having properties for treating or preventing disease in animals;
or
(b) Any substance or combination of substances which may be used in or administered to animals with a view either to restoring, correcting or modifying physiological functions by exerting a pharmacological, immunological or metabolic action, or to making a medical diagnosis.'

The relevant elements of the definition of a drug taken from the US FDC Act are as follows:

articles intended for use in the diagnosis, cure, mitigation, treatment, or prevention of disease in man or other animals;
and articles (other than food) intended to affect the structure or any function of the body of man or other animals.

Devices

EU legislation provides the following general definition of a medical device:

"medical device" means any instrument, apparatus, appliance, software, material or other article, whether used alone or in combination, together with any accessories, including the software intended by its manufacturer to be used specifically for diagnostic and/or therapeutic purposes and necessary for its proper application, intended by the manufacturer to be used for human beings for the purpose of:

- diagnosis, prevention, monitoring, treatment or alleviation of disease,
- diagnosis, monitoring, treatment, alleviation of or compensation for an injury or handicap,
- investigation, replacement or modification of the anatomy or of a physiological process,
- control of conception,
and which does not achieve its principal intended action in or on the human body by pharmacological, immunological or metabolic means, but which may be assisted in its function by such means.

Figure 1.4 Legal definitions of drugs and devices.

The subcategories, active implantable medical devices and in vitro medical devices are further defined as:

'active implantable medical device' means any active medical device which is intended to be totally or partially introduced, surgically or medically, into the human body or by medical intervention into a natural orifice, and which is intended to remain after the procedure

'in vitro diagnostic medical device' means any medical device which is a reagent, reagent product, calibrator, control material, kit, instrument, apparatus, equipment, or system, whether used alone or in combination, intended by the manufacturer to be used in vitro for the examination of specimens, including blood and tissue donations, derived from the human body, solely or principally for the purpose of providing information:
— concerning a physiological or pathological state, or
— concerning a congenital abnormality, or
— to determine the safety and compatibility with potential recipients, or
— to monitor therapeutic measures.

The US FD&C Act just provides the following general definition of a device:

The term device means an instrument, apparatus, implement, machine, contrivance, implant, in vitro reagent, or other similar or related article, including any component, part, or accessory, which is—

Intended for use in the diagnosis of disease or other conditions, or in the cure, mitigation, treatment, or prevention of disease, in man or other animals,

or

Intended to affect the structure or any function of the body of man or other animals,

and which does not achieve its primary intended purposes through chemical action within or on the body of man or other animals and which is not dependent upon being metabolised for the achievement of its primary intended purposes.

Figure 1.4 (*Continued*)

- An *asthma inhaler* is an example of a product that contains both a drug and a drug delivery device. Such a product would be regulated primarily as a drug as it achieves its medical purpose by pharmaceutical means. The inhaler would additionally have to satisfy the requirements of a device.

- *Stents* used to stabilise damaged arteries are often supplied impregnated with anti-clotting or other drugs. Such products are more likely to be regulated as *devices* as their primary purpose or mode of action is to provide a structural support for the artery. However, the drug could not be used in the device without marketing authorisation under the drug regulations.

- A *breath test,* used to determine the presence of *Helicobacter pylori,* associated with stomach ulcers, is an example of a diagnostic product involving separate drug and device components. To perform the test, a patient swallows some isotopically (e.g. ^{13}C or ^{14}C) labelled urea, which is then metabolised by the organism, releasing CO_2. A sample of the breath is taken and analysed for the presence of labelled CO_2. The sampling kit consists of the labelled urea, which is a drug; a sampling straw, which is a device; and a sample container, which would be considered an *in-vitro* diagnostic (IVD) medical device under EU definitions. Other examples of diagnostic drug products used in conjunction with medical devices include dyes administered to visualise blocked veins and arteries.

- *Ultrasound* and *X-ray equipment* are examples of diagnostic medical devices. *In-vitro* medical devices are distinguished from other diagnostic medical devices, in that a specimen must first of all be removed from the donor. A device worn by a diabetic that continually monitors their glucose via a non-invasive method (near-infra-red energy emissions) would be just regulated as a medical device, whereas a glucose-monitoring device that used a lancet to obtain a blood sample would be an IVD.

- Finally, a *test kit for analysing specimens without a medical purpose* would fall outside the regulations. For example, a test for therapeutic drug monitoring would be regulated as an IVD. However, a test could use the same technology for detecting a drug of abuse, but would be outside the scope of the regulations if it were only supplied for forensic testing.

1.6
Chapter Review

In this chapter, you learned that safety, efficacy and quality are key elements in attaining the ultimate goal of regulation – that of the protection of public health. The chapter explored the process by which legislation is introduced and identified the core legal texts that define the requirements for marketing drugs and devices. Finally, the chapter examined the legal definitions of drugs and devices and provided examples of how these can be applied to a selection of products.

1.7
Further Reading

- Legislative process
  http://www.fsai.ie/legislation/irish_and_eu/index.aspIrish and EU Legislation
- EU Directives and Regulations
  http://ec.europa.eu/enterprise/pharmaceuticals/eudralex/index.htm
  http://ec.europa.eu/enterprise/medical_devices/legislation_en.htm
- EU Pharmaceutical legislation on CD
  http://ec.europa.eu/enterprise/pharmaceuticals/eudralex/homecd.htm

- Guidance on demarcation between medical devices and medicines
  MEDDEV Guide 2.1/3
  MEDDEV Guide 2.14/1
  http://ec.europa.eu/enterprise/medical_devices/meddev/index.htm
- US Legislation
  www.access.gpo.gov
  www.fda.gov
  www.aphis.usda.gov

2
Regulatory Strategy

2.1
Chapter Introduction

Chapter 1 examined the aims of regulation and identified the basic regulations currently in existence. This chapter looks at the broad strategies that are employed to achieve these goals, and the agencies that participate in leading industry towards these goals.

2.2
Basic Regulatory Strategy

The basic regulatory strategy employed to safeguard public health relies on focussing attention on three main areas of activity: product development; product manufacture; and market vigilance (see Figure 2.1).

2.2.1
Product Development

Before a product can be marketed, product developers must generate sufficient data to demonstrate that the new products they wish to bring to market are safe and effective. In the case of all drugs and "high-risk" medical devices these data must be presented to the regulatory authorities for review. If the review is satisfactory, the regulatory authorities will grant a marketing authorisation to enable commercial sale to commence. The regulatory authorities are also involved at this stage in approving any trials required to assess safety and effectiveness in human subjects (clinical trials) and in ensuring that quality systems and standards are applied where appropriate to the manufacture and evaluation of the materials in this pre-market phase.

2.2.2
Product Manufacture

Once commercial production commences, the main thrust of regulatory endeavour comes to bear on the manufacturing sites. All manufacturing sites must be registered

Medical Product Regulatory Affairs. John J. Tobin and Gary Walsh
Copyright © 2008 WILEY-VCH Verlag GmbH & Co. KGaA, Weinheim
ISBN: 978-3-527-31877-3

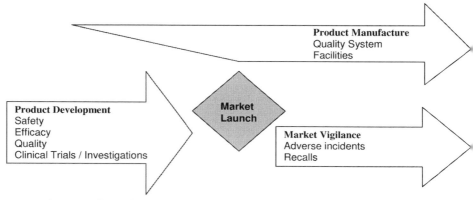

Figure 2.1 The regulatory strategy.

with the authorities, and the sites can then expect regular visits from inspection bodies. The purpose of such audits is to verify that manufacture is conducted under hygienic conditions in suitable facilities, and that appropriate quality systems are applied to control the process.

2.2.3
Market Vigilance

The final element in a regulatory strategy is market vigilance over the lifetime of the product. The regulatory authorities put in place systems so that all adverse incidents encountered in the market-place can be reported. The manufacturer is also required to have a system for capturing such incidents, and is obliged to report them to the regulatory authorities, even if the incident occurred in a market outside of that regulated by the authority. In this way, risks that did not surface during the pre-marketing phase can be identified, allowing for appropriate remedial action to be taken in a timely fashion. The flow of information as applied to vigilance systems in the European Union (EU) is illustrated in Figure 2.2.

2.3
Quality Assurance Systems

The application of quality assurance systems and standards is a fundamental element of the strategy to ensure that products are consistently safe, effective and of acceptable quality. Different quality systems are stipulated depending on the activities involved, such as good manufacturing practice (GMP), good laboratory practice (GLP), good clinical practice, or more general quality assurance systems for design, development, production, installation and servicing (e.g. ISO 9001). The specific aspects of the different systems will be described in later chapters of this book, but at this stage it is appropriate to examine some of common core principles on which the systems are based.

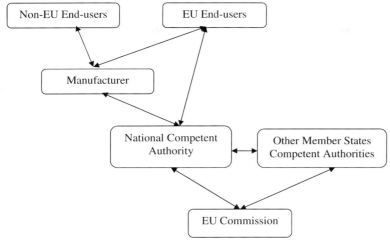

Figure 2.2 EU vigilance communication systems.

2.3.1
Personnel

People are a key element in all quality systems. Every single individual in an organisation must understand what their responsibilities are, and how the task that they perform can affect quality. Of course they must also be capable of doing the job. Job descriptions should be created to define the duties, responsibilities and authority level of each position in an organisation. Job specifications are used to identify the qualifications, skills and experience that are required. Review of an individual's qualifications, skills and experience in comparison to the job description will identify the training needs for that individual. A qualified engineer may have all the knowledge necessary for an engineering position, but will still have basic training requirements in terms of company policies, procedures, software systems, and so on. For many companies, a generic training plan is prepared in the form of an induction programme that can be tailored to an individuals needs. Records of all training delivered should be documented and maintained in a training record that can be made available to auditors for examination during an inspection (see Figure 2.3).

The *overall management structure* must be clearly outlined in organisational charts or other appropriate means. To demonstrate the organisation's commitment to quality, it is normal for the Quality Manager to report directly to the Chief Executive Officer. Regular

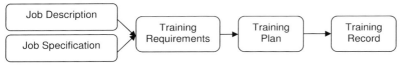

Figure 2.3 The training process.

management reviews of the quality system are also an essential element and again it is common practice for all the senior management of a facility to participate in such reviews.

2.3.2
Documentation

Documentation is another essential ingredient in all quality systems. It serves as a mantra for auditors and inspectors that ". . . if it is not documented it does not exist or it never happened". The documentation requirements can be split into two main categories: (i) procedures outlining what must be done; and (ii) records showing what has been done.

The performance of all tasks must be described in detailed, written procedures. Procedures describing recurring tasks are usually referred to as Standard Operating Procedures (SOPs) or General Operating Procedures (GOPs), whereas procedures describing non-repetitive tasks are called protocols (e.g. clinical trial protocol). Specifications complement procedures and are used to set the standards that apply to any stage of a product or process.

A document control function is required to regulate the authorisation and circulation of such documents. This involves the sign-off on a master copy of the document by authorized personnel, followed by a recorded circulation of official copies to relevant departments. If a document needs to be updated, the revised version should be reviewed and approved by the same functions that approved the original. The revised version is then issued with a new version number and date, and all copies of the obsolete version should be destroyed other than the obsolete master, which may be kept for historical traceability.

Records of all tasks performed and results obtained should be written down in black or blue ink, if using paper records. Pencil is not acceptable. If you discover a mistake you should draw a single line through the error and write the correct information above or beside it, accompanied by the date and your initials (see Figure 2.4). You should never obliterate the error with correcting fluid. Your records could end up as evidence in legal proceedings. Records in the form of various types of technical files serve to satisfy authorities that the products that you developed are safe and effective, and may be authorized for marketing. While such dossiers may just contain condensed data summaries it is vital that the original raw data records can be produced if necessary.

Traditionally, documentation has been maintained on hardcopy paper systems. However, soft copy computer records are permitted provided it can be assured that the records can be maintained without corruption. More sophisticated systems that are used to create electronic records and electronic signature must be capable of

153 JJT 20/4/06
The average was calculated as ~~135~~ This is ....

Figure 2.4 Correction of an error.

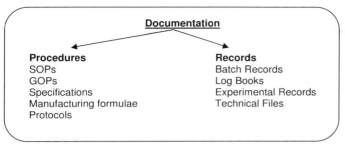

Figure 2.5 Documentation summary.

identifying the time, date and person responsible for creating or amending the record. Specific regulations dealing with these requirements will be discussed further in Chapter 11. Figure 2.5 provides an overview of the types of documents associated with most quality systems.

2.3.3
Facilities and Equipment

Quality systems require that facilities and equipment should be appropriate to the activities undertaken. Surfaces that are easy to clean and maintain in hygienic condition are a requirement in many situations. For example, cloth-backed chairs would not be acceptable in a laboratory that handled potentially biohazardous materials. Equipment should be checked at installation to demonstrate that it can perform its desired function. This is frequently done using an Installation Qualification, Operational Qualification and Performance Qualification (IQ/OQ/PQ) commissioning process. Routine maintenance and calibration programmes are then required to ensure that equipment continues to deliver the specified performance.

2.3.4
Corrective and Preventive Action

Quality systems are intended to be self-correcting and self-improving. Triggers that prompt corrective or preventive actions can come from a number of sources.

The regular internal auditing of a quality system is a fundamental requirement. These audits should be conducted by personnel independent of the operational area being audited; for example a quality department should not audit its own activities. The results of such audits, together with those from outside inspection bodies, will invariably identify areas for improvement.

Feedback from the market is another key data source. This can come through customer comments or complaints or through the vigilance systems that are mandatory for drugs and devices.

Data collected on the day-to-day operation of a facility are a more immediate source of information. This requires the recording of all non-conformances, procedure deviations, product failures, or product recalls. A variety of statistical tools may be used to analyse the data and identify underlying trends.

Thorough investigations should be conducted according to a predetermined SOP to ensure that the root cause is identified, if not already obvious. Once this has been done, a corrective action plan must be agreed and the process closed out by verifying that the corrective or preventative action has been implemented effectively.

2.4
Validation

Validation is an activity that can be defined as *providing confirmation by examination and objective evidence that the particular requirements for a specific intended use can be consistently fulfilled.*

It can be applied to either a product design or a process. It is important for you to appreciate the difference between verification and validation.

Verification means *confirmation by examination or direct measurement that specified requirements are met.*

The following examples may serve to illustrate the differences between both activities.

If you were a box manufacturer producing boxes of specified dimensions, it could be practically feasible to measure each box individually and verify that it meets the specification. On this basis, you could have complete confidence that each box will be capable of holding objects of a certain size.

However, there are many situations where verification studies may be either not practical or not possible so as to assure that a product or process functions as required. Consider a process where you produce packaged sterile devices. You could only verify that an individual device is sterile by breaking the seal and testing, after which it is no longer sterile – that is, the verification testing is destructive. Validation studies offer the only means of providing a high level of confidence that all the devices will be sterile. The validation studies in this case would involve challenging the sterilisation process by exposing the devices to a significant microbiological load, and then verifying that sterilization is achieved under the minimum sterilisation condition. Once you can demonstrate that the process works under worst-case conditions, you can then have confidence that it will work consistently under normal controlled operating conditions.

For product validation, the objective is to demonstrate that the product will meet the users' needs when operated under the intended use conditions. Consider the previous example of the box. The dimensional verification studies would have shown that it was capable of holding objects of a certain dimension. However, this alone would not guarantee that the box could perform its intended containment and protection functions. The weight of the object and consequent handling requirements would have a significant bearing on the required rigidity of the box. Similarly, if it was intended for use in cold storage, uptake of moisture from the surrounding damp conditions could have a negative impact on its rigidity. Field trials or testing under simulated-use conditions are often the only way to demonstrate that a product will function correctly under the totality of conditions that may be experienced in normal use.

Validation also applies to software. In a simple example, you could create an Excel spreadsheet template with fixed formulae to calculate the mean and standard deviation of a range of data. To validate this template you would enter a set of sample data and verify the template-calculated results against the manually calculated results. In order to be confident that the template could be used for further data sets you would password-protect the cell formulae and verify that they cannot be altered without it.

When looking at these examples, you can see that verification is very much a retrospective activity, where the objective is to confirm that what has happened is correct. In contrast, validation takes a prospective view, with the objective being to establish confidence that a product or process will work satisfactorily going forward.

2.5
Regulatory Bodies

Having looked at the strategies that are used to achieve regulatory goals, we shall now consider the principal bodies involved in implementing these strategies.

2.5.1
European Commission

Executive responsibility for drugs and devices at European Commission level is located within the Directorate-General for Enterprise and Industry. This is the equivalent of a civil service department at national level. As such, the focus is more on the development of policy and legal/administrative provisions.

Separate units exist to deal with pharmaceuticals and devices. Their general policy aims are to:

- ensure a high level of protection of public health;
- promote a single market for drugs and devices;
- foster a stable and predictable environment for innovation.

The pharmaceutical unit is the more active of the units as there is much greater use of EU-centralised procedures in drug regulation compared to devices. However, the activities of both units can be broken down under four main headings: legislation; support and guidance; external relations; and information technology.

In the legislative arena the tasks of the units are to:

- maintain, update and simplify EU legislation whenever feasible;
- draft new legislation;
- propose decisions for the authorisation and surveillance of medicinal products under an EU centralised procedure (pharmaceuticals unit);

- propose Commission regulations on maximum residue limits (MRLs) of veterinary medicinal products in foodstuffs of animal origin (pharmaceuticals unit).

In supporting consistent application of the regulations at national level the units:

- draw up detailed guidance documents;
- provide guidance on legislation and ensure that it is properly implemented within the EU;
- support the mutual recognition of national marketing authorisation decisions for drugs (pharmaceuticals unit).

In external relations with non-EU countries the units:

- participate in international assemblies that promote global harmonisation of requirements;
- negotiate mutual recognition agreements with third countries;
- pursue cooperation with Central and Eastern European Countries;
- prepare for and verify the implementation of EC legislation in EU accession candidate countries.

In the area of information technology (IT) and communication, the units coordinate the development of IT systems that permit greater visibility and interaction between all concerned parties, whether they be regulators, manufacturers, or consumers. For pharmaceuticals in particular, an EudraNet system has been developed which permits the secure transfer, sharing and tracking of information between the regulatory bodies. It also hosts a number of databases including Community registers of medical products centrally approved for human and veterinary use, and the EudraLex collection of the legislation and guidance documents that constitute "The Rules Governing Medicinal Products in the European Union". This is a 10-volume compilation that is also available on CD.

The commission is assisted by a number of consultative committees and expert working groups. The directives provide for committees with formal consultative and regulatory functions. These contain representatives from each Member State and the Commission, and enable the states to express their opinion at the stage when new legislative measures or decisions are being contemplated or prepared. Through the use of IT, much of their business is conducted remotely without the need for formal meetings. The relevant committees are:

- The Standing Committee on Medicinal Products for Human Use
- The Standing Committee on Veterinary Medicinal Products
- The Committee on Medical Devices
- The Committee on Standards and Technical Regulations.

In addition to the formal committees specified in directives, the commission is assisted by other expert committees and working groups. Depending on the subject matter, these can be drawn from the regulatory bodies, the scientific community, or industry and end-user associations.

2.5.2
The EMEA

The second body with a pan-European regulatory remit is the European Medicines Agency (EMEA). Whereas, the European Commission is concerned mainly with legislative and administrative tasks, the EMEA focuses on providing supporting scientific expertise and advice. The Agency was established in 1995 as the European Medicines Evaluation Agency as a consequence of EU regulation No. 2309/93. This regulation created the first centralised Community procedures for the authorization of medicines. The name of the agency was shortened by a subsequent regulation to its current title of the European Medicines Agency, but it still uses the acronym EMEA. As outlined in Figure 2.6, the agency performs a number of tasks in pursuance of its mission statement. However, its most important roles can be summarised as the coordination of the scientific assessment of marketing authorisation applications that are made via the EU-centralised procedure and the provision of expert scientific advice to both regulators and industry.

The Agency is a decentralised body which is located in London, and consists of:

- an executive director;
- a board of management;
- a secretariat; and
- Scientific Committees.

Mission Statement
The mission of the European Medicines Agency is to foster scientific excellence in the evaluation and supervision of medicines, for the benefit of public and animal health.

Principal activities
Working with the Member States and the European Commission as partners in a European medicines network, the European Medicines Agency:
- provides independent, science-based recommendations on the quality, safety and efficacy of medicines, and on more general issues relevant to public and animal health that involve medicines;
- applies efficient and transparent evaluation procedures to help bring new medicines to the market by means of a single, EU-wide marketing authorisation granted by the European Commission;
- implements measures for continuously supervising the quality, safety and efficacy of authorised medicines to ensure that their benefits outweigh their risks;
- provides scientific advice and incentives to stimulate the development and improve the availability of innovative new medicines;
- recommends safe limits for residues of veterinary medicines used in food-producing animals, for the establishment of maximum residue limits by the European Commission;
- involves representatives of patients, healthcare professionals and other stakeholders in its work, to facilitate dialogue on issues of common interest;
- publishes impartial and comprehensible information about medicines and their use;
- develops best practice for medicines evaluation and supervision in Europe, and contributes alongside the Member States and the European Commission to the harmonisation of regulatory standards at the international level.

Figure 2.6 The EMEA mission statement and principal activities (reproduced with permission of EMEA).

The board of management are responsible for the general supervision of the Agency, and among other duties adopts the annual work programmes and budgets for the Agency. The board is comprised of one representative of each Member State, two representatives of the Commission, two representatives of the European Parliament, two representatives of patient groups, one representative of doctors' associations, and one representative of veterinarians' associations. Each board serves a term of 3 years, and is led by a chairperson elected from among its members.

The executive director is appointed by the board for a term of 5 years and is, together with a secretariat of approximately 360 staff, responsible for the day-to-day operation of the Agency. As such, the secretariat provide scientific, technical and administrative support to the Scientific Committees. The organisational structure of the secretariat is shown in Figure 2.7.

The Scientific Committees are key to the functioning of the agency. They are the scientific decision-making component responsible for delivering opinions and advice on any issue relating to medicines. There are currently five committees, each focussed on different areas:

- The Committee for Medicinal Products for Human Use (CHMP)
- The Committee for Medicinal Products for Veterinary Use (CVMP)
- The Committee on Orphan Medicinal Products (COMP)
- The Committee on Herbal Medicinal Products (HMPC)
- The Paediatric Committee (PDCO)

A further committee, The Committee for Advance Therapies, is to be established under regulation (EC) No. 1394/2007 to deal with issues raised by emerging drug products based on gene therapy, somatic cell therapy and tissue engineering.

Member states can nominate one expert and an alternative to each committee. These in effect come from the national regulatory bodies, which they also represent. The committees may additionally co-opt up to five other members. The committees can draw on the expertise of some 3500 scientists using a network that extends through the national regulatory bodies, and so form further advisory committees and working groups as appropriate.

2.5.3
National Competent Authorities

Executive responsibility for the administration of regulations at national level is divested in dedicated agencies. Although distinct from government departments, the Ministers of Health are the usual political masters of these national Competent Authorities. For example, the authority in Ireland is known as the Irish Medicines Board (IMB) and it covers both drugs and devices. Similar to the EMEA, the authorities are scientific and technical in orientation and are guided by various advisory panels. Their main tasks include:

- Evaluation of applications for drug marketing authorisations made either directly to the national authority or indirectly via the centralised procedure to the EMEA

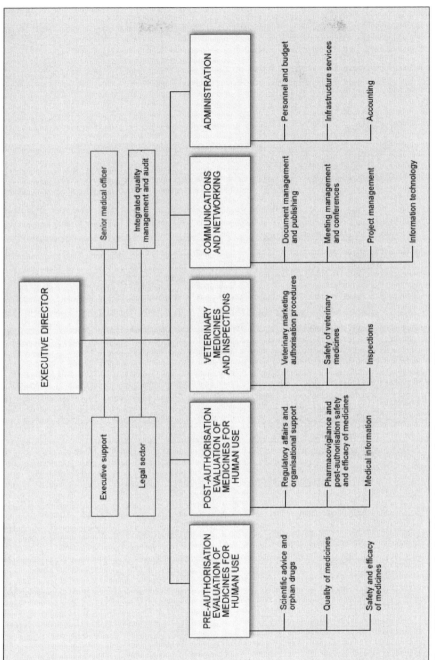

Figure 2.7 EMEA organisational chart (reproduced with permission of EMEA).

- Monitoring and approval of information provided on leaflets and promotional material
- National licensing of drugs
- Inspection and licensing of facilities
- Approval and licensing of clinical trials
- Operation of market vigilance systems at national level.

2.5.4
Notified Bodies

Notified Bodies play a special role in the regulation of medical devices. They are commercial inspection and auditing organisations rather than statutory bodies. They must be able to demonstrate that they have the technical competence and independence to deliver objective assessments. Although their accreditation status is designated by the member state where they are located, they can offer their services outside their state borders. They, rather than the Competent Authorities, provide the main interface with regulatory bodies for medical device manufacturers. By using Notified Bodies the Competent Authorities do not have to maintain the variety of specialist staff that would be required to assess the different types of technology platforms that are used for medical devices. Individual Notified Bodies do not have to be able to cover all specialist areas, but can just be designated as competent for specific types of device. For example, a Notified Body could just focus on assessing *in-vitro* diagnostic devices that have no electronic components.

2.5.5
The FDA

The Food and Drug Administration (FDA) is the principal regulatory agency responsible for food, drugs, devices and cosmetics in the USA. It has its origins in a Division of Chemistry that was established in the Department of Agriculture in the 1860s. The initial activities focussed on the adulteration of food with unsafe preservatives and colorants, and the general poor sanitary conditions that prevailed in food processing at that time. The Division of Chemistry became the Bureau of Chemistry in 1901 and continued to develop both its research and regulatory functions. In 1927 the Bureau was re-structured into two separate entities, with the regulatory function now located in the Food, Drug and Insecticide Administration. Its title was shortened to the Food and Drug Administration in 1930. Today, the FDA is a branch of the Department of Health and Human Services (HSS), and is headed by a Commissioner who is appointed by the President with the consent of the Senate. With over 9000 employees it is the largest regulatory authority in the world, and exercises considerable power in protecting the health of the US population. Key milestones in the development of the FDA and its regulatory functions are shown in Figure 2.8.

1862	**Bureau of Chemistry** established as part of the US Department of Agriculture.
1902	The **Biologics Control Act** is passed to ensure purity and safety of serums, vaccines, and similar products used to prevent or treat diseases in humans.
1906	The original **Food and Drugs Act** is passed prohibiting misbranded and adulterated food and drugs.
1912	**Sherley Amendments** prohibit labelling medicines with false therapeutic claims intended to defraud the purchaser.
1927	The Bureau of Chemistry is re-organised with regulatory functions now residing in the **Food, Drug and Insecticide Administration.**
1930	The name of the Food, Drug and Insecticide Administration is shortened to the **Food and Drug Administration** (FDA).
1937	An un-safely formulated sulfanilamide preparation kills 107 people, mainly children.
1938	The **Food, Drug and Cosmetic Act** is passed. This requires drugs to be shown to be safe before marketing, and authorises inspections of facilities.
1962	**Kefauver-Harris Drug Amendments** are passed. These require drug manufacturers to prove the effectiveness of new drugs.
1966	FDA contracts with the National Academy of Sciences/National Research Council to evaluate the effectiveness of 4000 drugs approved on the basis of safety alone between 1938 and 1962.
1972	**Over-the-Counter Drug Review** begun to enhance the safety, effectiveness and appropriate labelling of drugs sold without prescription.
1976	**Medical Device Amendments** passed to ensure safety and effectiveness of medical devices, including diagnostic products.
1983	**Orphan Drug Act** passed, enabling FDA to promote research and marketing of drugs needed for treating rare diseases.
1988	**Food and Drug Administration Act** officially establishes FDA as an agency of the Department of Health and Human Services with a Commissioner of Food and Drugs appointed by the President.
1992	The **Prescription Drug User Fee Act** allows the FDA to charge fees in order to hire more staff to speed up the review process and thus deliver faster approval decisions.
1996	The **Quality System Regulation** is introduced for medical devices, requiring developers of high-risk devices to apply design controls.

Figure 2.8 Milestones in the development of regulatory oversight of drugs and devices in the USA.

In pursuit of its mission of protecting public health, the FDA performs a wide variety of tasks that include the following:

- Development of regulatory policy;
- Establishment of regulations and guidance document in support of the FDC Act;
- Review and approval of pre-marketing submissions for drugs and devices;

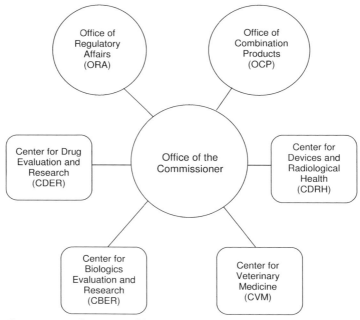

Figure 2.9 Partial structure of the FDA.

- Inspection of facilities and products with follow-up enforcement actions, where necessary;
- Ensuring that labelling, packaging and promotional material is truthful, informative and not misleading; and
- Participation in international initiatives on global harmonisation.

The FDA is organised into a number of offices and centres, the most relevant of which in the context of drugs and devices are shown in Figure 2.9. The Office of the Commissioner acts as the administrative headquarters of the agency. The Office of Regulatory Affairs (ORA) is responsible for ensuring compliance with the regulations, and contains about one-third of the agency's personnel stationed in over 160 locations throughout the US. Its staff carry out regular audits of facilities to ensure that they operate to appropriate GMPs, and inspects and tests products for compliance with relevant standards.

Responsibility for the review and assessment of pre-marketing submissions for new drugs and devices is divided between four specialist centres.

- The Center for Drug Evaluation and Research (CDER) is the largest centre, with about 1800 staff. It is responsible for the evaluation of submissions made in pursuit of marketing authorisations for all chemical-based drugs for human use, together with some biological-based drugs that came within its remit in 2003 (see Table 2.1).

- The Center for Biologics Evaluation and Research (CBER) is responsible for the oversight of what might be considered more traditional biological products, such as

Table 2.1 Biological products regulated by the CDER and CBER.

CDER-regulated	CBER-regulated
Cytokines (e.g., interferons and interleukins)	Blood and blood components (e.g., albumin, clotting factors)
Therapeutic enzymes	Vaccines
Thrombolytic agents	Antitoxins, venoms and antivenins
Hormones	Cell- and tissue-based products (e.g., bone marrow, corneas, but not vascularised organs, such as heart, kidneys, lungs)
Monoclonal antibodies for *in-vivo* use	
Growth factors	Polyclonal antibodies
Additional miscellaneous proteins	Gene therapy products
	Allergenic extracts

blood and blood products, vaccines, toxins and antitoxins. It is also responsible for the evaluation of new drug products that are emerging from cutting-edge biomedical research on human gene therapy, cell and tissue transplants, xenotransplantation (transplants from animals), transgenic plants and animals and proteomics, genomics and bioinformatics. As part of its mandate to ensure the safety of products of human origin, the CBER is responsible for medical devices and diagnostic tests used in the collection, screening and processing of blood and tissue products. The split between the CDER and the CBER in terms of the biological products that each regulates is shown in Table 2.1.

- The evaluation of chemical-based veterinary medicines rests with the Center for Veterinary Medicine (CVM).

- General medical devices are the responsibility of the Center for Devices and Radiological Health (CDRH). The CDRH is also responsible for assessing the safety of non-medical radiation-emitting products such as televisions, microwaves and mobile telephones.

Some products may have features that do not fit neatly under the remit of a single centre (e.g. drug-eluting vascular stents). Hence, to assist such product developers the Office of Combination Products (OCP) was established in 2002. The OCP can decide as to which centre to assign the role of lead assessor, and is also active in reviewing existing inter-centre agreements and in fresh initiatives to facilitate a smoother regulatory framework for such products.

2.5.6
US Department of Agriculture

The US Department of Agriculture (USDA) has regulatory responsibility for veterinary biologicals, which is exercised by its Animal and Plant Health Inspection Service (APHIS). The regulated products include vaccines, bacterins and diagnostics, which are used to prevent, treat, or diagnose animal diseases. The Center for Veterinary

Biologics (CVB), which is located within APHIS, is responsible for the review and licensing of such products. As outlined above, the CVM of the FDA regulates chemical-based veterinary medicines.

2.5.7
Pharmacopoeia Authorities

Pharmacopoeia organisations are not directly involved in the regulation of the pharmaceutical industry, but they play an important supporting role by developing pharmacopoeia standards and methods and providing reference materials for standardisation. The European Pharmacopoeia is produced by the European Directorate for the Quality of Medicines and Healthcare (EDQM), which has its headquarters in Strasbourg. The EDQM operates under the authority of the Council for Europe, which has a broader base than the European Union in that it contains 37 members, including all 27 EU states. Members are obliged to take the necessary measures to adopt European Pharmacopoeia standards as their national standards. The EDQM also coordinates quality assurance standards and proficiency testing schemes for national Official Medicines Control Laboratories (OMCLs), cooperates in harmonisation activities with US and Japanese pharmacopoeia organisations, and provides leadership in the area of tissue transplantation and blood transfusion. Its interactions with other European authorities are illustrated in Figure 2.10. The United States Pharmacopeia (USP) is the equivalent official public standards-setting authority for all prescription and over-the-counter medicines, dietary supplements and other healthcare products manufactured and sold in the United States.

2.6
International Harmonisation Bodies

Regulatory bodies across the world share a common goal of protecting public health. However, as regulatory systems were developed independently in each country, many differences in specific requirements emerged. It is well recognised that such differences pose a barrier to the introduction of new products in terms of both cost and time. Additionally, they create a difficult environment for what is a significantly multinational industry to operate in. To address this deficiency a number of harmonisation bodies have been created, initiated in the first instance through the United Nations World Health Organization (UN WHO).

2.7
International Conference on Harmonisation

The International Conference on Harmonisation (ICH) of technical requirements for the registration of pharmaceuticals for human use was established in 1990 as a joint regulatory/industry project to improve, through harmonisation, the efficiency

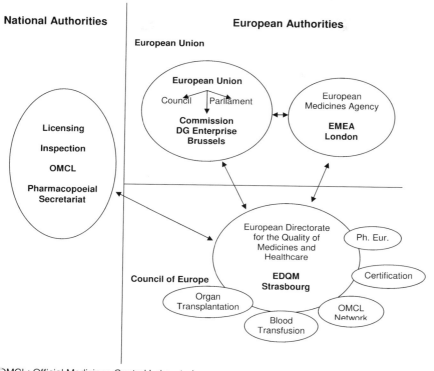

National Authorities

European Authorities

European Union

OMCL: Official Medicines Control Laboratories
Ph. Eur.: European Pharmacopoeia
Certification: Certification of suitability of Monographs of the European Pharmacopoeia

Figure 2.10 Illustration of the interaction of the EDQM with the overall regulatory system for medicines in Europe. Reproduced with permission of EDQM, ©EDQM, Council of Europe.

of the process for developing and registering new medicinal products in Europe, Japan and the United States. The ICH brings together the regulatory authorities from the USA, the EU and Japan, along with experts from pharmaceutical industry trade associations in the three regions. The respective representative bodies from each region are shown in Figure 2.11.

Harmonisation activities are overseen by a steering committee comprising representatives of each of the six parties, together with observers from the WHO, the European Free Trade Association (EFTA) and Canada. The Committee meet at least twice each year, with the location rotating between the three regions. The committee also organises major international conferences on harmonisation that are held every 2 years.

Most of the practical work on harmonisation is undertaken by topic-specific Expert Working Groups (EWGs), which again include representatives from each of the six participating groups. Much of the work is carried out remotely with the aid of communications technology, but meetings are also held to coincide with the steering

Regulators
The European Commission (EC)
The US Food and Drug Administration (FDA)
The Japanese Ministry for Health, Labour and Welfare (MWHLW)

Industry Representatives
The European Federation of Pharmaceutical Industries and Associations (EFPIA)
The Pharmaceutical Research and Manufacturers of America (PhRMA)
The Japanese Pharmaceutical Manufacturers Association (JPMA)

Figure 2.11 The ICH composition.

committee meetings. The steering committees and EWGs are supported by a secretariat provided by the International Federation of Pharmaceutical Manufacturers Association (IFPMA).

The ICH focuses on developing harmonised guidance documents under four topic headings:

- "Quality" (Q) addressing chemical and pharmaceutical quality assurance.
- "Safety" (S) relating to *in-vitro* and *in-vivo* pre-clinical studies.
- "Efficacy" (E) dealing with clinical studies in human subjects.
- "Multidisciplinary" (M) covering the format of authorisation applications, and so on.

A five-step process is used to develop new guidance documents (see Figure 2.12). The process is initiated by presenting a concept paper to the Steering Committee.

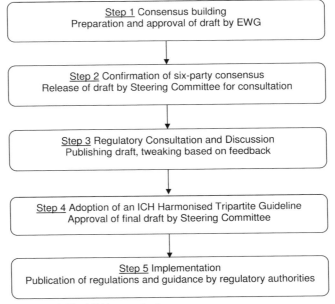

Step 1 Consensus building
Preparation and approval of draft by EWG

Step 2 Confirmation of six-party consensus
Release of draft by Steering Committee for consultation

Step 3 Regulatory Consultation and Discussion
Publishing draft, tweaking based on feedback

Step 4 Adoption of an ICH Harmonised Tripartite Guideline
Approval of final draft by Steering Committee

Step 5 Implementation
Publication of regulations and guidance by regulatory authorities

Figure 2.12 The ICH harmonisation process.

This should outline existing differences and suggest how harmonisation on a particular topic could be achieved. If the proposal is accepted, the Steering Committee assigns the topic to a new or an existing EWG, as appropriate.

1. The first step is consensus building, where the EWG keep working at the task until they reach a scientific consensus and produce a draft document. Drafting of the document is left to a single "rapporteur", who adjusts the document based on the comments and feedback from the parties.

2. The process then moves to step 2, the confirmation of the six-party consensus. This requires agreement from the Steering Committee, which then "signs off" the draft and releases it to the three regulatory authorities.

3. In step 3 – regulatory consultation and discussion – the draft is published by each of the regional regulatory authorities for public consultation. If necessary, a second "rapporteur", from one of the regulatory parties, is appointed to collate the feedback from the regulatory authorities. The rapporteur will bring the document through the same consensus process as step 1 until the feedback from the regions has been incorporated to the satisfaction of the EWG.

4. The final draft is then presented to the Steering Committee in step 4 for adoption as an ICH Harmonised Tripartite Guideline.

5. In the final step of implementation, the regional regulatory authorities must issue the guideline as regulations or official guidance, as appropriate to their regulatory system. One of the key factors in the success of the ICH is the commitment from the regulatory authorities to implement guidelines that have come through the harmonisation process.

2.7.1
VICH

A parallel body, the International Conference on Harmonisation of Technical Requirements for the Registration of Veterinary Medicinal Products (VICH), came into being in 1995 to work on the harmonisation of veterinary product regulations. It has a similar steering committee and working group structure to the ICH, but uses a nine-step process for developing or revising guidelines, as outlined in Figure 2.13. Its secretariat operates from the Office International des Epizooties (OIE) in Paris.

2.7.2
The Global Harmonisation Task Force

The Global Harmonisation Task Force (GHTF) was conceived in 1992 to address similar issues for medical devices. It has a broader regional base than its pharmaceutical counterparts in that it includes representatives from Canada and Australia in its core group, in addition to those from the EU, US and Japan. In some ways, it also faces a stiffer challenge in that there is more divergence in the regulatory

Step 1
The Steering Committee defines a priority item from a concept paper prepared by a member and appoints a working group if needed. A topic leader is given a mandate to draft a recommendation.

Step 2
The expert working group drafts a recommendation.

Step 3
The draft is submitted to the Steering Committee for approval of its release for consultation.

Step 4
Following adoption by the Steering Committee, the draft recommendation is circulated for consultation.

Step 5
The working group takes comments into consideration in preparing a revised draft. The topic leader must be a representative of a regulatory authority at this stage.

Step 6
The revised draft recommendation is submitted to the Steering Committee for approval.

Step 7
A final recommendation and proposed implementation date are circulated to the relevant regulatory authorities.

Step 8
Steering Committee members report back on the implementation progress in their regions.

Step 9
Recommendations may be revised at the request of a member to take into account new scientific evidence.

Figure 2.13 The VICH harmonisation process.

approaches that are used and in the range of medical devices that exist. Its organisational structures were slow to evolve compared to the ICH, and it was only in 2001 that a Steering Committee was instituted with responsibility for management, oversight and policy setting for the GHTF. It operates four study groups, which focus on developing harmonised guidance in the subject areas outlined in Figure 2.14.

2.8
Pharmaceutical Inspection Cooperation Scheme

The Pharmaceutical Inspection Cooperation Scheme (PICS) was established in 1995. It brings together regulatory authorities from different countries, for the purpose of developing harmonised GMP requirements and inspection techniques, with the goal

Study Group 1
Definition of a medical device
Risk-based classification of devices
Essential principles of safety and performance
Labelling
The role of standards
Clinical evaluations
Summary technical files

Study Group 2
Adverse event reporting
Vigilance report forms
Handling information concerning vigilance

Study Group 3
Quality systems for design and manufacture
Process validation
Design control process

Study group 4
Regulatory auditing

Figure 2.14 Topics for the GHTF study groups.

of achieving confidence in, and mutual recognition of, GMP inspections carried out by individual member regulatory authorities. It emerged from an earlier Pharmaceutical Inspection Convention, which was established by EFTA members in 1970. However, as this was a treaty-based convention, it was no longer possible for EU Member States to participate in mutual recognition conventions at national level, once the European legislation addressed intra-community and external recognition obligations. As part of their remit, the PICS provides training for inspectors and organises conference to discuss emerging GMP issues. The countries of the participating regulatory authorities are shown in Table 2.2.

Table 2.2 Participants of the Pharmaceutical Inspection Cooperation Scheme (PICS) by country.

Austria	Denmark	Finland	Latvia
Liechtenstein	Spain	Portugal	Sweden
Greece	United Kingdom	Hungary	Ireland
Romania	Germany	Italy	Belgium
France	Netherlands	Slovac Republic	Czech Republic
Poland	Norway	Iceland	Switzerland
Canada	Australia	Malaysia	Singapore
Estonia	Malta	South Africa	Argentina

The present list of PIC/S participating authorities may have been updated. For the latest up-to-date list, see www.picscheme.org.

2.9
The World Health Organization (WHO)

The United Nations World Health Organization is active in promoting coordinated responses to health issues at a global level. It is particularly concerned with health issues in developing countries, and plays a significant role in ensuring that emerging diseases, vaccine problems or adverse drug reactions are adequately communicated outside of first-world countries.

2.10
Chapter Review

This chapter has examined the basic regulatory strategy that is deployed to protect public health against the hazards of unsafe drugs and devices. It also looked at some common principles of quality systems and how they serve as a basic tool in implementing a regulatory strategy. Finally, the chapter reviewed the structure and remit of the main players charged with applying the regulations in Europe and the US, and the efforts to achieve global harmonisation. In the following chapters you will experience how the regulatory strategy works in practice as new products are brought to the market.

2.11
Further Reading

- European Commission, Enterprise and Industry Directorate-General
  http://ec.europa.eu/enterprise/pharmaceuticals/index_en.htm.
  http://ec.europa.eu/enterprise/medical_devices/index_en.htm.
- European Medicines Agency
  http://www.emea.europa.eu/htms/aboutus/emeaoverview.htm.
- Irish Medicines Board
  www.imb.ie.
- FDA
  www.fda.gov.
- USDA APHIS
  www.aphis.usda.gov.
- ICH
  www.ich.org.
- VICH
  www.vichsec.org.
- EDQM
  www.edqm.eu.
- USP
  www.usp.org.
- GHTF   www.ghtf.org.

3
Drug Discovery and Development

3.1
Chapter Introduction

Leading pharmaceutical companies are reliant on the discovery of new drugs to maintain their revenue streams, as the manufacture of established medicines tends to migrate to lower-cost operations, once patent protection has expired. As you will discover over the following chapters, the overall process of bringing a new drug to market is complex, time-consuming and expensive. Typically, the process can take an average of 10 to 12 years to complete and cost in excess of €450 million. Furthermore, the failure rate is quite high, and for each successful drug that does emerge up to 10 000 potential candidates may have been examined at the outset. Before studying the specific regulatory requirements that govern the introduction of a new drug to the market, this chapter sets out some of the basic principles that apply to the categorisation, discovery and development of drugs and drug products.

3.2
Drug Categorisation

There are a number of ways in which drugs may be categorised, depending on the criteria applied. It is worthwhile introducing some of these now as it can provide a context to help your understand the later text.

3.2.1
Prescription Status

Drugs are classified from a regulatory perspective, as either over-the-counter (OTC) or prescription-only medicines (POM, or Rx in America). Human medicines are subject to medical prescription where they:

- are likely to present a danger either directly or indirectly, even when used correctly, if utilized without medical supervision; or

Medical Product Regulatory Affairs. John J. Tobin and Gary Walsh
Copyright © 2008 WILEY-VCH Verlag GmbH & Co. KGaA, Weinheim
ISBN: 978-3-527-31877-3

- are frequently and to a very wide extent used incorrectly, and as a result are likely to present a direct or indirect danger to human health; or
- contain substances or preparations thereof, the activity and/or adverse reactions of which require further investigation; or
- are normally prescribed by a doctor to be administered parenterally (i.e. by injection).

Human prescription drugs may be further categorised into:

- those available on *renewable* prescription;
- those *not* available on renewable prescription;
- those available *only* on special prescription; and
- those available on *restricted* prescription.

The non-renewable prescription status may be applied to drugs that are potentially addictive, or drugs such as cortisone, that can cause damage with extended usage. The "restricted use" category may be applied to certain classes of antibiotics that are used to treat infections such as tuberculosis, where it is vital to ensure that the organism does not have the opportunity to develop resistance. OTC drugs may also be distinguished depending on whether they are available in general retail outlets (e.g. supermarkets) or only through pharmacies.

New drugs will only be authorised as prescription medicines in the first instance, as it requires considerable experience in the market to convince the regulators that it is safe to reclassify a drug as an OTC product. The term "new drug" may mean a new pharmaceutical substance, or a new therapeutic application of an existing one.

3.2.2
Physical Properties

Drugs may also be categorised on the basis of their physical properties as either small- or large-molecule categories. Small-molecule drugs can be based on simple inorganic compounds such as the antacid, sodium bicarbonate, or lithium, which is used in the treatment of manic depression. However, most drugs are organic molecules of varying complexity, and although many were originally sourced from natural materials the majority are now produced via synthetic organic chemistry. Large-molecule drugs are predominantly protein-based substances and generally referred to as *biopharmaceuticals*. Rather than using a synthetic route, they are produced either by extraction from a biological source or, more commonly, by a combination of genetic engineering and cell/tissue culture techniques.

3.2.3
Mode of Action

Another way to categorise drugs is on the basis of their mode of action. Some drugs function simply by replacing a naturally occurring substance that is deficient or defective in a particular disease or condition. Examples of this include insulin to treat diabetes, blood-clotting factors to treat haemophilia, growth hormone to treat dwarfism, and hormone replacement therapy to treat acute menopause. Diagnostic drugs may depend on just their physical presence or particular properties. Barium was used as an electron-dense material to facilitate stomach X-ray studies, but is now largely redundant since the development of endoscopes. The *Helicobacter pylori* breath test depends on the use of a substrate containing an isotope of carbon. If the bacterium is present in the stomach, it metabolises the substrate, releasing isotopic carbon dioxide, which can be detected in a sample of the subject's breath shortly after ingestion of the drug.

However, most drugs rely on a binding interaction with an appropriate target to achieve their effect. All biological processes are governed by binding interactions at the molecular level. Enzymes depend on the binding of a substrate to the active site in order to achieve their catalytic effect. The regulation of processes is achieved by binding of various messenger molecules to a diverse range of receptors that trigger either stimulatory or inhibitory responses. Our immune defence mechanisms depend on the creation of antibodies which are able to bind to molecules that are seen as foreign to the body. Many drugs function by being able to mimic the substances that normally interact with the binding sites of molecules that are critical to a process. The effect may be to inhibit an enzyme or to elicit either a stimulatory (*agonist*) or inhibitory (*antagonist*) response in a receptor. Today, the identification of relevant enzymes and receptors, and the molecules that can interact to produce a desirable response, is the key to most drug discovery activities.

3.2.4
Therapeutic Use

Drugs may also be classified on the basis of their therapeutic use. In Table 3.1 are listed the top 15 therapeutic categories as represented by the total number of products at various stages in the regulatory approval process in the United States in 2006. General therapeutic categories can be broken down into different sub-categories depending on the various classification schemes that may be used.

The World Health Organization (WHO) promotes the use of an Anatomical Therapeutic Chemical (ATC) classification system for the collection and analysis of data on drug use. This was originally developed by Scandinavian authorities, and uses a combination of anatomical, therapeutic and chemical criteria to assign drugs to an individual class. The top-level categories, which are anatomically based, are listed in Table 3.2.

3.3
Drug Discovery

Many early drug discoveries came from simple observations and experimentation with naturally occurring products. For example, the discovery of the first vaccine

Table 3.1 The top 15 therapeutic categories on the basis of new products in the regulatory process in the United States in 2006.

General therapeutic category	Example(s)
Cancer	Cytokines
Central nervous system (CNS)	Antidepressants
Cardiovascular system	Antihypertensives
Infections	Antibiotics
Pain/inflammation	Analgesics
Respiratory	Anti-asthmatics
Diabetes	Insulin
Gastrointestinal	Anti-ulcerants
Haematology	Clotting factors
Vaccines	MMR, poliomyelitis
Arthritis	
Dermatology	
Autoimmune disease	
Ophthalmology	
Bone metabolism	

came from the studies which Jenner conducted with cowpox to protect children against the more lethal chickenpox. Similarly, the discovery of penicillin came from observations by Fleming of the inhibitory effect of penicillin mould on the growth of bacteria. As a fledgling pharmaceutical industry emerged, drug companies undertook more organised searches in which various biological specimens were screened for potential drug compounds. Although the biological world contains a vast array of novel bioactive molecules, the chances of success from such a crude randomised

Table 3.2 The first level of the ATC system.

Code	Anatomical system
A	Alimentary tract and metabolism
B	Blood and blood-forming organs
C	Cardiovascular system
D	Dermatologicals
G	Genito-urinary system and sex hormones
H	Systemic hormonal preparations, excluding sex hormones and insulins
J	Anti-infectives for systemic use
L	Antineoplastic and immunomodulating agents
M	Musculoskeletal system
N	Nervous system
P	Antiparasitic products, insecticides and repellents
R	Respiratory system
S	Sensory organs
V	Various

approach are low. However, these odds can be improved by adopting a more focused screening approach. Initially, a widely used practice was to examine the traditional cures and herbal remedies of indigenous people, in the hope of identifying active pharmaceutical ingredients with therapeutic value. Similarly, plants that were resistant to disease or predation were targeted, as were animals and insects known to produce toxic compounds. This approach has led to the discovery of many of the drugs that are still in use today. Moreover, the potential still exists to discover drugs by this strategy, as it is estimated that less than 1% of plant species have been screened to date. More recently, however, drug discovery has come to rely on a knowledge-based approach in which an understanding of the disease process at a molecular level provides the key.

The foundation for the more rational approach to drug discovery that is practised today comes from basic research into disease processes and medical conditions. Much of this research is carried out in universities and other research institutions, with funding coming from both government and pharmaceutical industry sponsors. By understanding a disease or a condition at the molecular level, it then becomes possible to identify potential targets that could be exploited to achieve a desired therapeutic effect.

This point is well illustrated by the following examples:

- *AIDS* After considerable research, the causative agent for AIDS was identified as a virus, which was eventually named the Human Immunodeficiency Virus (HIV). The virus was characterised as a retrovirus – which means that it carries its genetic code in the form of RNA and requires a reverse transcriptase enzyme to generate the DNA copies that are necessary for replication. This process of reverse transcription, which is unique to retroviruses, was therefore seen as a prime candidate for therapeutic intervention. The drug azidothymidine (AZT; zidovudine) already existed, and had been investigated as a possible anticancer drug. The thesis was that, because it has a similar structure to the DNA nucleotide thymidine, but lacks a functional group that enables further propagation of a DNA sequence, it might be capable of slowing down the DNA replication activity which is one of the features of rapidly growing tumours. Although, unfortunately, AZT did not prove effective when tested, it was re-examined when a means to interfere with the replication of the HIV virus was being sought. Fortunately, in this case, AZT was able to fool the reverse transcriptase enzyme into incorporating it into the DNA copy instead of thymidine, the effect being to prevent further propagation of such DNA chains. However, although AZT did display a significant inhibitory effect, it alone was not sufficient to prevent replication of the virus. Investigation of the replication process identified another step that could be targeted. The viral proteins are translated as large molecules that require cleavage by a protease before they can be assembled into new viral particles. Inhibitors of this protease were identified as providing a second means of attacking the replication process. Finally, direct inhibitors of the reverse transcriptase enzyme were developed, so that today's medical practitioners are able to effectively combat AIDS by using a combinational therapy strategy. Similar examples of how the understanding of a disease process or condition can lead to improved drug therapies include those of peptic ulcer and hepatitis B.

- *Peptic ulcers* Peptic ulcers are caused by the presence of excess stomach acid which attacks the lining of the stomach, and which if left untreated, could lead to perforation of the gut and haemorrhage, resulting in death in extreme cases. The earliest drugs used to treat the condition included simple antacids that worked by just neutralising the acid. By the early 1960s it was known that histamine was able to stimulate the secretion of gastric acid. Thus began a search to identify structural analogues of histamine that could block what was then a hypothesised new histamine receptor (H_2-receptor) involved in the acid secretion pathway, as opposed to other histamine receptors associated with allergic responses. This culminated with the launch in the 1970s of cimetidine (brand name Tagamet), which was the first of a new class of drugs for the treatment of ulcers, termed H_2 antagonists. This was quickly followed by the introduction of another structural analogue, ranitidine (brand name Zantac) in 1981, which displayed significantly reduced side effects compared to cimetidine. Research then focused on identifying inhibitors of the proton pumps, which are responsible for actively transporting H^+ across the cell membranes into the stomach against a concentration gradient. The first proton-pump inhibitor, omeprazole (brand names Losec or Prilosec) was introduced in the mid-1980s. Proton-pump inhibitors quickly superseded H_2 antagonists as the more effective ulcer treatments because: (i) they are non-competitive inhibitors which covalently and irreversibly block the proton pumps, whereas the H_2 antagonists are just competitive inhibitors that can be displaced by histamine; and (ii) they target the actual terminal step in the generation of acid in the stomach. Other research noted the presence of the bacterium *Helicobacter pylori* in the stomachs of the majority of patients that suffered from ulcers. These bacteria can survive in the gastric mucosal layer that acts as a buffer between the acid environment of the stomach and the epithelial cells that surround it. It is believed that, similar to the natural action of food in the stomach, they stimulate gastric cells to release the hormone gastrin, which in turn binds to receptors on the cells that produce gastric acid (parietal cells) resulting in increased acid secretion. Today, the recommended treatment for recurrent ulcers consists of a combination of antibiotics (amoxicillin, clarithromycin) to eradicate the *H. pylori* and thus eliminate the root cause of the ulcer and a proton-pump inhibitor (omeprazole) to temporarily stop acid secretion so as to allow the ulcer to heal. The various intervention points in the treatment of ulcers are illustrated in Figure 3.1.

- *Hepatitis B* Hepatitis B vaccines provide another illustration of how drug products have advanced with increasing technical capability. Vaccination against hepatitis B is common practice for health workers, travellers and others who may be at risk of exposure to the virus. The initial vaccines contained inactivated virus to promote the immune response necessary to protect against future infection by the live virus. However, there was always some concern in case there was not complete inactivation of the virus used for vaccination. Further research into the virus identified the surface proteins against which the immune response is raised. The genetic

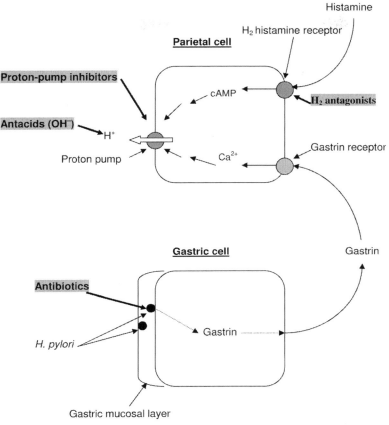

Figure 3.1 A schematic representation of the control mechanism that stimulates gastric acid secretion, and the intervention points used to treat ulcers. The parietal cells and gastric cells form part of the epithelial cell lining of the stomach. Histamine release is usually triggered as part of the enteric nervous system response to distension of the stomach when food is eaten. Interaction of the food with the gastric mucosal layer is the normal trigger for gastric cells to release gastrin, which is then carried by the bloodstream to the parietal cells. Calcium ions and cyclic AMP act as intracellular messengers in the transfer of the signal from the receptors to the proton pumps of parietal cells where the acid is generated.

sequence for these proteins has been isolated and genetically engineered into "safe" organisms for expression. Thus, today's vaccines are much safer as they contain only the antigenic surface proteins of the hepatitis B virus.

Having identified targets that have the potential to be manipulated to therapeutic effect, the next step is to identify suitable molecules that can interact with these targets. In a "blind" approach, large libraries of compounds of either natural or synthetic origin are screened for binding to the putative target. This requires the development of suitable assays to measure binding. These binding assays may be based on the measurement of a response that is triggered by binding, or on using compounds labelled in such a fashion that they can be detected; examples include dye, fluorescence, enzyme or radioactive ^{14}C-labelling methods.

In the rational approach, focus is placed on identifying the amino acid sequence and 3-D structure of potential binding sites, with the view of predicting the structure of molecules that will give a best-fit interaction with these sites. The determination of an accurate 3-D structure requires the application of X-ray crystallography or NMR spectroscopy to a purified sample of the potential target. Computer modelling and simulation programmes can then be used to "design" virtual molecules that should interact with the binding sites. Finally, it is up to the organic chemists actually to synthesise these molecules for testing.

Information technology is gaining an increasingly wider role in drug discovery, and the broad area of computer applications in biology has evolved into a specialised discipline known as *bioinformatics*. Databases have been generated that contain the DNA sequence of many organisms, including that of the human genome. From the DNA sequence, it is then possible to deduce the primary amino acid sequence of proteins, to predict their 3-D structures, and even to suggest functions for some of the proteins that remain to be isolated *in vitro*. At its current stage of development, the modelling may not be very exact, but as further data become available it can be expected that computer simulations will become more precise and useful in the discovery of new drugs.

Knowledge of the genetic code – and the control of its expression – is opening up new horizons for therapeutic intervention. Defects in specific genes, or malfunctions in their expression, have now been identified as the root cause for many non-infectious diseases. New biopharmaceutical drugs rely largely on the ability to isolate functional copies of the genes and to genetically engineer them into producer organisms. There is also considerable ongoing research activity in trying to replace defective genes *in vivo* through either gene therapy or cell therapy. The use of stem cells to create replacement tissues and organs is seen as a route for future advance. However, a number of ethical questions have been posed regarding the use of stem cells of embryonic origin, and these difficulties must be resolved before the practical benefits of this line of research can be realised. An overview of the drug discovery and development process is shown in Figure 3.2.

3.4
Drug Development

Once potential drug compounds have been identified, a long and arduous process of drug development and evaluation ensues. Before embarking on the wide variety of activities and studies that are required, it is normal practice to examine structurally related compounds, particularly in the case of small-molecule drugs. There are a number of reasons for this. First, it is possible that an analogous compound could actually be more potent than the candidate compound initially identified. Second, by examining a broad range of compounds, it reduces the

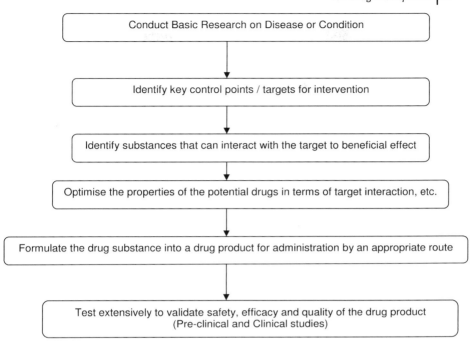

Figure 3.2 Modern drug discovery and development processes.

chances of a competitor being able to bring similar drugs to market. At this early stage, once initial compounds have been characterised and shown to have potential therapeutic value, it is advisable to prepare a *patent application*. Patents grant legal protection to applicants to exclusively commercialise their "invention" in exchange for making a detailed description of the "invention" available to the public. To obtain a patent, the subject of the invention must be novel, have utility, and should advance the "state of the art" beyond that which previously existed. The standard term of protection that is granted is 20 years from the date a patent application is filed. Drug developers may apply for an extension of the protection period to take account of time lost while conducting clinical trials and obtaining marketing authorisations. The standard maximum extension permitted is five years, subject to the restriction that it does not result in a total protection period post-authorisation of more than 15 years in Europe or 14 years in the US. Specific paediatric studies can add a further six months' protection.

This protection provides a vital incentive to drug development companies, as patented medicines are usually able to command a premium price – the presumption being that the new drug is more effective than any existing alternative therapies. The patent holder may also gain revenue by charging a licensing fee to other companies that wish to manufacture the drug or use it as part of a combination therapy. In this way, the developer has an opportunity to recoup the substantial costs that go into drug development.

3.5
Drug Delivery

A key element in the development process is to select and optimise the most appropriate method for the delivery of the drug. Drugs may be administered by:

- injection;
- ingestion;
- inhalation; or
- application to dermal or other accessible body surfaces.

The development of the most appropriate route for administration is influenced by a number of factors, including the location of the target within the body, the desired speed and duration of the therapeutic effect and the properties and characteristics of the drug substance.

3.5.1
Location

Obviously, if you wish to treat a skin condition or infection, a preparation that can be applied topically would be the preferred option. Similarly, inhalation would be the first choice if trying to treat a pulmonary or bronchial condition, such as asthma. Dermal application would also be the first choice for localized tissue treatments (e.g. muscle injury), provided that the drug can be absorbed through the skin. However, in most other situations it is necessary for drugs to enter the bloodstream in order for them to be transported to their site of action. This is most commonly achieved by ingestion, or by intravenous (i.v.), intramuscular (i.m.) or subcutaneous (s.c.) injection when the oral route is not suitable.

3.5.2
Drug Characteristics

The characteristics and properties of the drug will have a big bearing on its ability to enter the body and reach its target, particularly with respect to its size and relative solubility in polar versus non-polar solvents. All body fluids are water-based, and are thus compatible with polar solutes (hydrophilic substances). Cells on the other hand, are surrounded by cell membranes, which are based on a lipid bilayer structure that creates a non-polar core that is not very compatible with polar solutes. Excluding injection, drugs must be able to pass through cellular tissues to enter the body. Small-molecule drugs with significant non-polar character (hydrophobic substances) are able to diffuse passively through membranes. More polar or charged molecules will have greater difficulty in passing through the lipid bilayer, but may cross cell membranes with the assistance of appropriate transmembrane proteins that provide channels for passive diffusion or active transport. These mechanisms are illustrated schematically in Figure 3.3. Relative polarity will also have a significant bearing on the

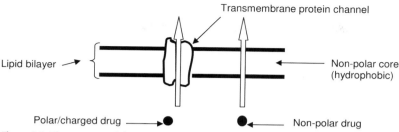

Figure 3.3 The passage of drugs across cell membranes.

solubility of the drug, and may dictate the level of "free" drug circulating in the bloodstream, and the distribution pattern between various tissues. A non-polar drug will display slower dissolution rates and solubility in aqueous solutions, such as are encountered in the gastrointestinal tract, while most of the drug will be bound to proteins such as albumin when circulating in the bloodstream. This type of drug is also likely to show a tendency to deposit in fat cells, which could reduce the effectiveness of a therapy in obese people. Chemists may have the opportunity to manipulate the polarity of the drug substance during development so as to favourably influence its behaviour. *Hydrophobic character* is determined by the amounts of non-polar carbon–carbon and carbon–hydrogen bonds that are present in a molecule. *Polarity/water solubility* can be improved by introducing groups such as hydroxyl, carboxyl, amino, sulphate or phosphate into a compound. Thus, the drug developer may be able to optimise the properties of a drug substance by adding chemical groups at locations remote from the site that is involved in interaction with the target.

Large-molecule drugs have greater difficulty in traversing cell membranes, and to date most biopharmaceutical products are administered via injection (i.e. parenterally), either intravenously (i.v.), intramuscularly (i.m.), or subcutaneously (s.c.). However, as any diabetes sufferer will attest, this is not the most pleasant prospect when frequent dosing is required, and drug companies are very keen to develop alternative routes for the administration of biopharmaceuticals. Although large-molecule drugs have great difficulty in crossing the cell membranes, they can enter many cells by the process of *pinocytosis*, whereby the substances are enveloped within a portion of the cell membrane to form an *endosome* (see Figure 3.4). Despite this occurrence, ingestion is not a viable option for biopharmaceuticals, as they will face denaturation by the acid environment of the stomach, and degradation by digestive enzymes in the gut. Adsorption via the lungs has shown greater promise, and the first such product (an inhalable insulin, Exubera®) was approved for general medical use in 2006.

3.5.3
Speed and Duration of Therapeutic Effect

The desired speed and duration of a therapeutic effect will also influence the development of the final drug product. The standard treatment of meningitis with antibiotics is via intravenous administration so as to achieve an immediate effect,

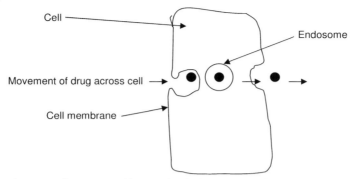

Figure 3.4 The passage of large-molecule drugs across cells.

while rapid pain relief is promoted as a feature of many common analgesic remedies, which are invariably manufactured for oral administration. On the other hand, the treatment of blood pressure or an irregular heartbeat requires reasonably constant levels of the drug to be achieved in the circulatory system. Under such circumstances, a slow-release form of the drug is required, normally as an oral dosage variant. As mentioned above, the properties of the drug substance will have a bearing on the ability to achieve this. The rate of release can be further controlled by the way in which the drug is mixed or formulated with other filler substances, termed *excipients*. A list of common excipients is provided in Table 3.3.

Ingestion remains the most popular route for drugs delivery due to its convenience and the high level of patient compliance generally attained. However, there are potential obstacles that the drug developer may need to address here. Although the acid environment in the stomach may prove detrimental to some drugs, this can be overcome by encapsulating the drug in a protective enteric coating that only allows it to be released when it reaches the intestine. A second disadvantage and one which may be less easily overcome – is that, once adsorbed through the intestine, the drugs are carried first to the liver, which is the principal organ for drug metabolism. Consequently, there is the risk that a considerable portion of the drug may lose its therapeutic activity before it reaches the site of action. This is termed the hepatic first-pass effect.

Table 3.3 Common excipients used in the formulation of drug products.

Magnesium stearate	Lactose
Microcrystalline cellulose	Starch (corn)
Silicon dioxide	Titanium dioxide
Stearic acid	Sodium starch glycolate
Gelatin	Talc
Sucrose	Calcium stearate
Pregelatinized starch	Hydroxypropyl methylcellulose
Hydroxypropyl cellulose	Ethylcellulose
Calcium phosphate	

In summary, the characteristics of the drug and the disease or condition may suggest a likely route and mode of delivery. Considerable work is required to develop and optimise an appropriate drug presentation. The suitability of any particular strategy must be verified by undertaking the extensive pharmacodynamic and pharmacokinetic studies that form part of the pre-clinical and clinical trials.

3.5.4
Stability

As discussed above, the stability of a drug as it enters the body can have a significant bearing on how it is presented. The inherent stability of the drug may also decide the final form in which the drug product is presented. Because of stability limitations, many biopharmaceutical products are presented in a freeze-dried form that is only reconstituted with water prior to use. The stability of solid tablet forms, and the type of packaging used, may be dictated by the tendency to absorb moisture. It may be possible to reduce this hygroscopic behaviour or improve stability by using different salts or combinations of excipients.

3.6
Chapter Review

This chapter examined various ways by which drugs may be categorised, for example by prescription status, mode of action, physical properties and therapeutic use. The basic strategies used in the discovery of drugs were also examined, ranging from simple observations to the collection of traditional cures and screening of animal and plant substances to today's research-based strategies that rely on gaining an understanding of disease processes at the molecular level. Finally, the chapter considered the drug development process in terms of choosing an appropriate drug form and method of delivery.

3.7
Suggested Reading

- AIDS History and Treatment
  www.avert.org.
- ATC Code
  www.whocc.no/atcddd/atcsystem.html.
  www.dkma.dk/visUKLSArtikel.asp?artikelID=7748.

4
Non-Clinical Studies

4.1
Chapter Introduction

The previous chapters examined the broad framework of regulations in terms of purpose, strategy and players, together with some of the general principles and considerations that are applied to the discovery and development of drugs. It is now time to examine how the regulatory requirements impact on the development of a new drug. Over the next three chapters, you will examine the essential steps that a developer must take from a regulatory perspective in order to bring a new drug for human use to market. The overall process can be broken down into three sequential stages of: (i) pre-clinical studies; (ii) clinical studies; and (iii) marketing authorisation. This chapter examines the non-clinical studies that are required.

4.2
Non-Clinical Study Objectives and Timing

The majority of non-clinical studies are undertaken in the pre-clinical phase of drug development. These studies serve both to guide the developer and satisfy the regulatory authorities. The objectives of this phase can be summarised as follows:

- To formulate the drug substance into a putative drug product that can be manufactured and administered to effect.
- To generate sufficient data so as to convince the regulators that it safe and worthwhile to proceed to clinical trials with human subjects.

Ultimately, the non-clinical data will form a substantial part of the regulatory submission that is required for marketing authorisation.

Safety and quality aspects are the main topics that must be addressed from a regulatory perspective at the pre-clinical phase of drug development. Indicative efficacy data will also be obtained, but authoritative data can be obtained only from clinical studies conducted with humans. Safety and preliminary efficacy indications

Medical Product Regulatory Affairs. John J. Tobin and Gary Walsh
Copyright © 2008 WILEY-VCH Verlag GmbH & Co. KGaA, Weinheim
ISBN: 978-3-527-31877-3

Safety	Quality
Pharmacology	**Chemistry, Manufacturing, Control**
• Pharmacodynamics • Pharmacokinetics • Pharmaceutical safety	• Chemical characterisation • Manufacturing process • Control of materials • Biotech product quality
Toxicology	**Stability**
• Single-dose toxicity • Repeated-dose toxicity • Genotoxicity • Carcinogenicity • Reproductive toxicity	• Substance stability • Product stability

Figure 4.1 Non-clinical study topics.

are obtained from investigations of the toxicology and pharmacology of the drug. Data on the composition (chemistry), manufacture, control and stability of the drug substance and drug product must be generated in order to satisfy quality concerns. These broad areas of work are summarised in Figure 4.1.

Many of these studies can run in parallel. However, ethics and cost considerations require that as much of the work as possible is carried out *in vitro* before moving on to *in-vivo* studies in laboratory animals. There is considerable interplay between the studies, and data must be continuously reviewed and assessed so that planned investigations can be adjusted in line with the most recent findings. Unfavourable results at any stage in the process could require taking a step back to an earlier stage or, indeed, abandoning the development completely. Some non-clinical studies can run into other development phases, as illustrated in Figure 4.2.

Extensive guidance is available from the ICH, EMEA and FDA websites. This outlines the expectations of the authorities in terms of data requirements, and provides approaches and rationales that may be used to develop a study programme. ICH Multidisciplinary guideline M3(R1) 'Non-Clinical Safety Studies for the Conduct of Human Clinical Trials for Pharmaceuticals' provides a good overview of the topic, as it presents an overview of the requirements and timing of non-clinical safety studies. The ICH harmonised safety and quality guidance documents, listed in Figure 4.3, should be consulted next, after which it may necessary to refer to non-harmonised guidance documents on the EMEA or FDA websites.

4.3
Pharmacological Studies

Pharmacology refers to the study of how drugs interact with living organisms and bring about a change of function. Investigations of the pharmacological properties

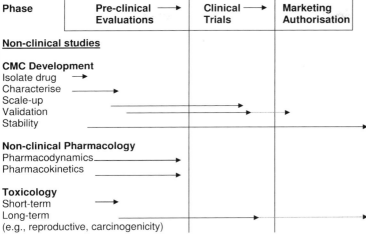

Figure 4.2 How some non-clinical activities can continue into the clinical trial and marketing authorisation phases.

and behaviour of the drug constitute a significant portion of the initial non-clinical studies that must be conducted. These can be broken down into *pharmacodynamic* and *pharmacokinetic* studies, using a combination of *in-vitro* and animal models.

4.3.1
Pharmacodynamic Studies

Pharmacodynamics is a discipline within the broader topic of pharmacology, which focuses on how a drug brings about a particular response, and the effective levels that are required in order to elicit such a response. Some of these basic data will have emerged from the research-based activities that initiate the development of most drugs today. However, considerable additional studies are required to establish detailed dose–response curves so that the optimum therapeutic level can be chosen.

In-vitro studies are of great practical benefit as they enable you to investigate drug/target interactions and responses, without the complicating pharmacokinetic effects that will be seen *in vivo*. Consider by way of example an enzyme that will be targeted by drug inhibitors. Once you have an assay that enables you to measure enzyme activity, you can determine the activity at different concentrations of substrate and then plot activity versus substrate concentration. Typically, this will yield either a hyperbolic or sigmoid response curve, as shown in Figure 4.4. A feature of all enzyme-catalysed reactions is that you approach a maximum rate of reaction, termed the V_{max}, which corresponds to the point at which all the active sites of the enzyme are saturated with substrate. The substrate concentration yielding 50% of this maximum activity is termed the K_m, and equates to the concentration at which 50% of the active sites of the enzyme are saturated with substrate. The lower this value, the greater the affinity of

Safety
Carcinogenicity Studies
S1A   Need for Carcinogenicity Studies of Pharmaceuticals
S1B   Testing for Carcinogenicity of Pharmaceuticals
S1C(R1)   Dose Selection for Carcinogenicity Studies of Pharmaceuticals & Limit Dose
S2A   Guidance on Specific Aspects of Regulatory Genotoxicity Tests for Pharmaceuticals
S2B   Genotoxicity: A Standard Battery for Genotoxicity Testing of Pharmaceuticals
Toxicokinetics and Pharmacokinetics
S3A   Note for Guidance on Toxicokinetics: The Assessment of Systemic Exposure in Toxicity
            Studies
S3B   Pharmacokinetics: Guidance for Repeated Dose Tissue Distribution Studies
Toxicity Testing
S4   Duration of Chronic Toxicity Testing in Animals (Rodent and Non-Rodent Toxicity Testing)

Reproductive Toxicology
S5(R2)   Detection of Toxicity to Reproduction for Medicinal Products & Toxicity to Male Fertility
Biotechnological Products
S6   Preclinical Safety Evaluation of Biotechnology-Derived Pharmaceuticals
Pharmacology Studies
S7A   Safety Pharmacology Studies for Human Pharmaceuticals
S7B   The Non-Clinical Evaluation of the Potential for Delayed Ventricular Repolarization (QT
            Interval Prolongation) by Human Pharmaceuticals
Immunotoxicology Studies
S8   Immunotoxicity Studies for Human Pharmaceuticals

Quality
Stability
Q1A(R2)   Stability Testing of New Drug Substances and Products
Q1B   Stability Testing: Photostability Testing of New Drug Substances and Products
Q1C   Stability Testing for New Dosage Forms
Q1D   Bracketing and Matrixing Designs for Stability Testing of New Drug Substances and
            Products
Q1E   Evaluation of Stability Data
Q1F   Stability Data Package for Registration Applications in Climatic Zones III and IV
Analytical Validation
Q2(R1)   Validation of Analytical Procedures: Text and Methodology
Impurities
Q3A(R2)   Impurities in New Drug Substances
Q3B(R2)   Impurities in New Drug Products
Q3C(R3)   Impurities: Guideline for Residual Solvents
Pharmacopoeias
Q4   Pharmacopoeias
Q4A   Pharmacopoeial Harmonisation
Q4B   Regulatory Acceptance of Analytical Procedures and/or Acceptance Criteria (RAAPAC)
Quality of Biotechnological Products
Q5A(R1)   Viral Safety Evaluation of Biotechnology Products Derived from Cell Lines of Human or
            Animal Origin
Q5B   Quality of Biotechnological Products: Analysis of the Expression Construct in Cells Used for
            Production of r-DNA-Derived Protein Products
Q5C   Quality of Biotechnological Products: Stability Testing of Biotechnological/Biological Products

Q5D   Derivation and Characterisation of Cell Substrates Used for Production of
            Biotechnological/Biological Products
Q5E   Comparability of Biotechnological/Biological Products Subject to Changes in their
            Manufacturing Process
Specifications
Q6A   Specifications: Test Procedures and Acceptance Criteria for New Drug Substances and New
            Drug Products: Chemical Substances (including Decision Trees)
Q6B   Specifications: Test Procedures and Acceptance Criteria for Biotechnological/Biological
            Products
Good Manufacturing Practice
Q7   Good Manufacturing Practice Guide for Active Pharmaceutical Ingredients
Pharmaceutical Development
Q8   Pharmaceutical Development
Quality Risk Management
Q9   Quality Risk Management

Figure 4.3 ICH harmonised safety and quality guidelines.

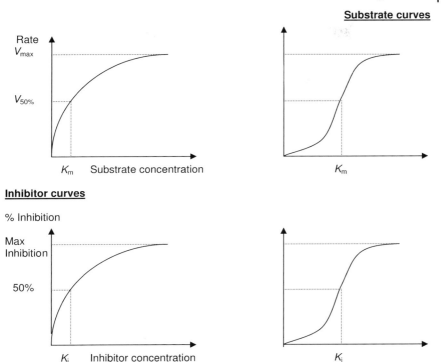

Figure 4.4 Binding curves of an enzyme for substrate and inhibitor.

the enzyme for its substrate. If you now proceed to measure enzyme activity at a fixed substrate concentration, but with varying concentrations of an inhibitor that competes with the substrate for binding to the active site, you will obtain similar dose–response curves, if this time you plot percentage inhibition versus inhibitor concentration. Again, you can identify the inhibitor concentration at which 50% inhibition is observed. This affinity constant is termed the K_i. The lower this value is, the less is required for significant binding, and thus the greater the potency of the inhibitor/drug.

A second aspect of relevance is the effectiveness or intensity of the response generated upon binding of the potential drug to its target. Continuing with the example of inhibition of an enzyme target, different types of inhibitors show different levels of inhibition, even when they have fully saturated their binding sites on the enzyme. *Competitive inhibitors* bind directly at the active site of the enzyme and, at saturation levels, can achieve almost total inhibition as the substrate is effectively blocked from binding to the active site of the enzyme. Other *non-competitive inhibitors* work by binding to sites outside the active site, and inducing conformational changes that reduce enzyme activity. The effectiveness of such inhibitors will depend on the extent to which they can induce conformational change but, even at saturation levels, it is less likely that they would be able to approach total inhibition.

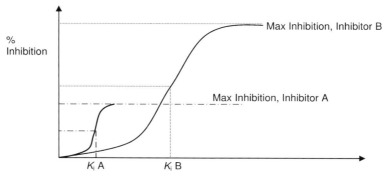

Figure 4.5 Illustration of the concepts of potency and effectiveness for two enzyme inhibitors. Inhibitor A has a strong affinity for the enzyme (low K_i), but even when it is maximally bound to the enzyme it only achieves partial inhibition – that is, it is potent but not very effective. Inhibitor B shows less affinity for the enzyme (high K_i), but it is a much more effective inhibitor when maximally bound – that is, it is not very potent but is effective.

The concepts of *potency* and *effectiveness* may be graphically illustrated by the dose–response curves shown in Figure 4.5. In essence, potency describes how well a drug binds to its target, whereas effectiveness is an expression of the magnitude of response that the drug can achieve once it is bound. Then, with an understanding of the dynamics of drug interaction at the molecular level, it is possible to proceed to dose–response studies in animal models, where pharmacokinetic and other factors come into play. Animal studies will also provide information on secondary pharmacodynamic effects, where the drug binds with targets other than the one intended, often with undesirable side effects.

4.3.2
Pharmacokinetic/Toxicokinetic Studies

Pharmacokinetics concerns the fate of a drug in the body at the approximate therapeutic dose range, while *toxicokinetics* assesses behaviour at the higher dose levels associated with toxic effects. The fate of a drug is dictated by the rates of:

- *Absorption* into the body;
- *Distribution* within the body
- *Metabolism* by the body; and
- *Excretion* from the body.

The effective concentration of drug that is available to interact with the target at the site of action is determined by the relative rates at which these processes proceed. Figure 4.6 illustrates the variation in the level of a drug that might be expected in the bloodstream over time following the ingestion of a single dose. A detailed knowledge of the processes governing the level of drug available to the target is thus essential in order to devise an effective drug product and therapy.

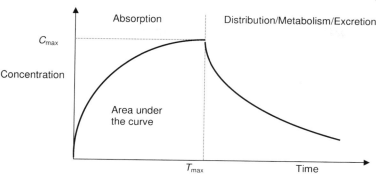

Figure 4.6 Time course of drug concentration in the bloodstream after a single dose.

As already mentioned, most drugs depend on entering the bloodstream for their transport to the site of action. Excluding intravenous injection, this will necessitate crossing cellular membranes. As discussed in Chapter 3, the rates of *absorption* are dependent on the physical properties of the drug substance and the way in which it is combined with excipients to form a drug product. When formulated correctly, many small-molecule drugs are capable of penetrating most cell membranes, including those of the dermis, whereas larger molecules are more restricted in their ability to enter the body – to date, entry via the nasal and pulmonary surfaces appears as the most promising. Careful measurement of the actual rates at which the drug is absorbed is the first stage in establishing its pharmacokinetic profile.

Once in the bloodstream, the drug will be carried to the various tissues where further *diffusion/absorption* take place. Again, the drug's properties may have a significant bearing on the pattern of distribution among the tissues. For example, non-polar drugs can diffuse through membranes with relative ease, which means that are more likely to be able to cross the blood–brain barrier. On the other hand, they may show an undesirable preference for fatty tissues, or the level of free drug in the blood may be too low to allow for effective diffusion into the target tissue. Serum albumin will act as a carrier for non-polar molecules, and if a drug has poor water solubility then most of it will be bound to albumin. A detailed knowledge of tissue distribution is essential in establishing the exposure levels for different organs and assessing the risks that this may pose.

As part of the process of elimination from the body, most drugs will undergo considerable *metabolism*. The liver is the major organ involved in drug metabolism. Cytochrome P450 enzymes play a key role in the metabolism of organic molecule drugs by introducing oxygen in the form of hydroxyl groups. This initiates a process of making them more water soluble, which can also include reduction, hydrolysis and the addition of polar conjugates. A detailed understanding of the metabolic fate of each drug is vital. For some drugs, the initial metabolism may yield a more pharmacologically active form of the substance, although in most cases the net effect is a loss of pharmacological activity. The toxicity of metabolites may also be of concern, as there is the possibility that some of the compounds formed could be significantly more toxic than the parent molecule. Additionally, many drugs will induce higher levels of cytochrome P450 enzymes, so that an increased rate of metabolism will be experienced, if the drug is to be taken over an extended period.

Excretion of the drug and/or its metabolites takes place primarily in the kidney, though some drugs also show considerable excretion via the bile/faeces. Again, it is important to study the rates of excretion of the drug or its metabolites, and to verify that there is no associated kidney damage.

In-vitro models can provide preliminary insights into some pharmacodynamic aspects. For example, cultured Caco 2 cell lines (derived from a human colorectal carcinoma) may be used to simulate intestinal absorption behaviour, while cultured hepatic cell lines are available for metabolic studies. However, a comprehensive understanding of the pharmacokinetic effects will require the use of *in-vivo* animal studies, where the drug levels in various tissues can be measured after different dosages and time intervals. Radioactively labelled drugs (carbon-14) may be used to facilitate detection. Animal model studies of human biopharmaceutical products may be compromised by immune responses that would not be expected when actually treating human subjects.

4.4
Bioavailability and Bioequivalence

Bioavailability studies play a critical role in the evaluation of product formulations throughout the entire development process.

Bioavailability is defined as the rate and extent to which the active ingredient or active moiety is absorbed from a drug product and becomes available at the site of action.

Bioavailability studies are used to compare different formulations of the drug product, or different batches of the same formulation and, as discussed in Chapter 8, generic copies of a reference drug. Their comparative value is based on the premise that, if similar amounts of identical active substance are delivered to the site of action at similar rates, then a similar biological response can be expected, which leads to the conclusion that the two preparations are *bioequivalent*.

In-vivo bioavailability studies are primarily based on measuring circulating levels of the drug substance in the bloodstream, as most drugs are ultimately delivered to the site of action via this route. In formulation development, the bioavailability resulting from the ingestion of different oral tablet presentations may be compared to direct injection, which represents the ideal 100% bioavailability target. If validated, the levels of drug or its metabolites in urine may be used as an indirect indicator of *in-vivo* bioavailability. For drugs that do not reach their site of action via the bloodstream, some other suitable biological indicator must be identified to monitor bioavailability.

In-vitro bioavailability tests usually form part of the criteria for evaluating the individual batches of a product. These are based on monitoring the rate of release of the drug substance from the pharmaceutical form, usually by observing the dissolution rates of tablets or capsules.

4.5
Toxicology Studies

Considerable studies are required to establish the toxicological profile of the drug substance. These must assess its direct toxic effects, together with its potential as a reproductive toxin, mutagen, or carcinogen.

4.5.1
Toxicity Studies

Most drugs can be expected to have potentially lethal effects, as their usefulness depends on their ability to influence physiological processes. However, the dose level at which toxic effects manifest should be significantly higher than those required to produce a pharmaceutical response, so that a good safety margin exists between therapeutic and toxic levels.

One of the first steps in investigating toxicity is to determine the level at which *acute lethal toxicity* is displayed. This is not only of relevance to the proposed therapeutic use of the drug but also provides essential data to assure the occupational health and safety of workers producing and handling the material. Acute toxicity is expressed in terms of a lethal dose $(LD)_{50}$ value, which is defined as the quantity of substance (in $mg\,kg^{-1}$ bodyweight) that is ingested or absorbed through skin, and that causes death within 14 days in 50% of the test animals. An LC_{50} value (concentration in air over 1 or 4 h) is used to express the toxicity of a substance that can be inhaled as a dust, mist vapour, or gas. LD_{50}/LC_{50} values are determined by administering different doses of the substance by ingestion, skin adsorption and inhalation to groups of test animals (rats or rabbits), and observing the number of fatalities occurring at each level within 14 days.

The LD_{50} studies may be integrated with the more extensive single-dose toxicity testing that is required as part of the toxicity evaluation of drug substances. Testing must be conducted in two mammalian species, with different drug levels adminis-tered via the proposed therapeutic route and intravenously, if different. Animals are observed for any evidence of toxic effect or changes in vital signs. All animals are then sacrificed and subjected to detailed post-mortem examination. Observations on some animals that displayed visible non-lethal effects may be continued past 14 days to see if the effects are reversible. These studies serve to identify organs that are subject to toxic effects, and also provide a basis for the selection of dose levels for the repeated dose studies that will follow.

Repeat-dose studies are conducted to evaluate the long-term effects of the drug. At least two species should be tested, one of which should be a non-rodent. The choice should include a species that produces a response close to that anticipated in humans. Three dose levels should be chosen, representing high, intermediate and low doses. The low level should correspond to the therapeutic range where no toxic effects are anticipated, while the high level should correspond to the range where mild toxic effects will manifest without encountering significant fatality. The drug will normally be given daily by the proposed route for administration in humans. The duration of the studies will depend on the proposed duration of the clinical trials in humans and the specific requirements of different regulatory authorities. A maximum duration of 6 months for rodents and 9 months for non-rodents has been agreed through the ICH process. The general monitoring of parameters such as body weight, clinical chemistry, haematology and ophthalmology is carried out over the course of the trial, and a detailed post-mortem examination of tissue and organs morphology is conducted at the end of the study. All results are referenced against a control group of animals.

Safety pharmacology studies may also be conducted in conjunction with the toxicity studies. These focus on identifying secondary pharmaceutical side effects that may occur, when the drug is administered in the therapeutic range. Emphasis is placed on identifying effects on vital systems, particularly the central nervous, cardiovascular and respiratory systems.

4.5.2
Genotoxicity Studies

Genotoxicity studies are required to identify compounds that can induce genetic damage ranging from single point gene mutations to gross alterations of chromosomal structure. Such effects are taken as indicative of the potential to cause cancer or heritable defects in humans. A standard battery of three types of test is recommended:

- A bacterial reverse-mutation test
- An *in-vitro* test with mammalian cells (chromosomal damage or mouse lymphoma tk assay)
- An *in-vivo* test in rodents (chromosomal damage in haematopoietic cells)

The principle of the bacterial reverse-mutation test is explained in Figure 4.7. The *in-vitro* tests may be conducted with and without metabolic activation with an extract of liver enzymes (as noted previously, drug metabolism occurs primarily in the liver and can result in metabolites that are more hazardous than the parent drug). The *in-vitro* testing will be performed before considering *in-vivo* tests, as they are much faster and cheaper to perform. The extent of testing may also be influenced by consideration of the structure of the substance and its similarity to known mutagenic compounds, or the degree to which it is absorbed by the body.

Bacterial reverse-mutation tests (also known as "Ames" tests) are based on using bacteria which carry a mutation that makes their growth dependent on a particular ingredient (usually an amino acid). The bacteria are inoculated onto culture plates that are deficient in the essential ingredient, and then exposed to different levels of the test substance. If the substance is mutagenic, it causes the bacteria to reverse-mutate and lose their dependence on the ingredient, thus enabling significant bacterial growth. If the substance is not mutagenic, then minimal growth will be observed.

Figure 4.7 An explanation of the bacterial reverse-mutation test (the Ames test).

4.5.3
Carcinogenicity Studies

Although positive results in genotoxic tests may be taken as being indicative of human carcinogenic potential, additional carcinogen testing may be required in order to identify other agents that can induce cancer through non-genotoxic mechanisms. The requirement for carcinogenicity studies is largely dependent on the proposed duration of treatment in humans. Trials should be performed where a minimum 6-month treatment is anticipated. A long-term study of up to 2 years in a single species (preferably rats) is now considered acceptable, with supplementary studies in other species, where necessary.

The drug should be administered at three levels by the route proposed for humans. The high dose level should be set so as to have relevance in humans. For drugs that display significant toxic effects, this may be related to the maximally tolerated dose in the toxicity tests, for example, the dose causing less than 10% deviation in body weight versus controls. If there is little evidence of toxicity it may be more appropriate to base the dose level on a multiple (usually 25-fold) of the maximum therapeutic dosage recommended in humans.

Carcinogenicity tests will usually not commence until preliminary clinical data are available to help with the study design. Marketing authorisation may be granted before the studies are completed. The requirement for such studies may be influenced by knowledge of how the drug is metabolised, and whether any of the metabolites are similar in structure to known carcinogens. Carcinogenicity tests are not normally required for protein-based drugs.

4.5.4
Reproductive Toxicology Studies

Reproductive tests are necessary to reveal any effects on mammalian reproduction. All stages ranging from pre-mating through conception, pregnancy and birth, to the growth of offspring are studied. Studies are conducted over a number of generations (preferably in rats), with different dose levels administered by the proposed route in humans. Considerable numbers of animals are required, as they must be autopsied at different stages in the reproductive/developmental process. An assessment of the data identifies any drugs that may result in impaired male or female fertility, developmental abnormalities in the foetus (teratogens), effects on lactation, or growth and maturation of the offspring.

4.6
Chemistry, Manufacturing and Control Development (CMC)

In addition to the safety studies, extensive investigations must be undertaken to establish the quality of the drug substance and product(s) during the pre-clinical phase. These will usually commence with drug discovery, where the initial characterisation of the substance occurs. This characterisation should be extended to gain as complete an understanding as possible of the structure, stereochemistry, impurities and chemical and physical characteristics and properties of the drug substance.

Initially, the drug will have been produced at the laboratory scale, and may have been extracted from a natural source. A vital element in the development process is to be able to scale up through pilot scale to commercial production. At a minimum, this will require optimisation of the manufacturing steps to yield a reproducible and robust process. In many cases this may involve developing a synthetic chemistry pathway to produce a substance that initially was extracted from a naturally occurring source, or by genetically engineering a biopharmaceutical into a suitable producer organism.

Analytical methods and specifications must be established and validated so as to define and control the quality and purity of the raw materials, intermediates and the finished product. For many standard chemical raw materials, the development of specifications will not be necessary as they are already published in US and European pharmacopoeia (for example, standards for water, organic solvents and various excipients). The ultimate objective of these activities is to be able to manufacture the drugs required for clinical trials in accordance with good manufacturing practice (GMP).

4.7
Quality of Biotech Products

Specific additional tests are required to assure the quality and safety of biopharmaceutical products derived from human or animal cell lines. There is always the possibility that harmful viruses such as herpesvirus may either be present in the original cell line, or could be introduced as adventitious agents in culture media during the growth of such cells. Thus, appropriate assays must be applied to detect such viruses and demonstrate that there is effective elimination or inactivation of any infective agents during the subsequent purification and processing steps. Considerable studies are also required to assess the stability of expression of products obtained from genetically modified organisms. Potential variation could arise because of the susceptibility to mutation of the primary DNA sequence, or alteration in the level of post-translational modification experienced under different culture conditions, particularly as a process is scaled-up. Specific guidance is provided in the ICH Q5 guidelines.

4.7.1
Stability Studies

Data to demonstrate the stability characteristics of both the drug substance and the drug product must be collected. Studies using three different batches of both substance and product in their respective containers/packaging must be conducted. Real-time data should be collected under conditions of temperature and relative humidity in line with the recommended storage. Conditions in different world climatic zones must be taken into consideration for cases where normal environmental storage conditions will be used. Standard conditions to simulate normal storage are: temperatures of 25 or 30 °C and relative humidity of 60 or 65%. A minimum of 12 months' real-time data is required for a marketing authorisation application. Real-time data should be supported by accelerated and intermediate stress testing at higher temperature/relative humidity. Photostability should be verified using a single batch.

4.8
Good Laboratory Practice (GLP)

All non-clinical studies used to demonstrate the safety of the drug must be conducted in accordance with the principles of GLP. Good laboratory practice is a quality system concerned with the organisational process and the conditions, under which non-clinical health and environmental safety studies are planned, performed, monitored, recorded, archived and reported. The principles of GLP were developed under the auspices of the Organisation for Economic Cooperation and Development (OECD) so that national regulators could mutually recognise and have confidence in safety studies that are conducted outside their jurisdiction. The GLP requirements are set out in EU legislation under Directive 2004/10/EC and in US regulations under 21CFR Part 58. The main headings of both regulations are outlined in Table 4.1. Although the headings are slightly different, the topics covered by the regulations are identical. It should be noted that the EU directive is also intended to cover safety testing of general chemicals and field studies.

As with all quality systems, the regulations address basic aspects such as personnel, documentation, facilities and equipment. However, there are some particular requirements that are worthy of note. A Study Director must be appointed with overall responsibility for the conduct and reporting of each study or group of related studies. A written protocol or study plan must be prepared outlining the purpose and detail of how the study will be conducted, with justification for approach taken where appropriate (duration, animals, etc.). An independent quality assurance (QA) group, in addition to the study director, must approve the study. The QA group are also responsible for oversight of the study, and must certify the final report with respect to GLP compliance. Samples of materials and tissues must be retained in addition to all documentary records. The retention time for samples and documents will depend on the stage of the authorisation process that the study supports and the requirements of the particular regulatory authority. Facilities must be inspected and accredited by external agencies; by notified bodies in the EU, or directly by the FDA in the US.

4.9
Chapter Review

This chapter examined the various non-clinical studies that are at the core of the pre-clinical phase of drug development. These studies are intended to gain a basic

Table 4.1 Comparison of the content of EU and US GLP regulations.

EU Directive 2004/10/EC	US 21 CFR Part 58
SECTION I INTRODUCTION Preface 1. Scope 2. Definitions of terms	**Subpart A – General Provisions** **Sec. 58.1 Scope.** **Sec. 58.3 Definitions.** Sec. 58.10 Applicability to studies performed under grants and contracts. Sec. 58.15 Inspection of a testing facility
SECTION II GOOD LABORATORY PRACTICE PRINCIPLES **1. Test facility organisation and personnel** 1.1. Test facility management's responsibilities 1.2. Study director's responsibilities 1.3. Principal investigator's responsibilities 1.4. Study personnel's responsibilities	**Subpart B – Organization and Personnel** Sec. 58.29 Personnel. Sec. 58.31 Testing facility management. Sec. 58.33 Study director. Sec. 58.35 Quality assurance unit. **Subpart C – Facilities** Sec. 58.41 General.
2. Quality assurance programme 2.1. General 2.2. Responsibilities of the quality assurance personnel	Sec. 58.43 Animal care facilities. Sec. 58.45 Animal supply facilities. Sec. 58.47 Facilities for handling test and control articles.
3. Facilities 3.1. General 3.2. Test system facilities 3.3. Facilities for handling test and reference items 3.4. Archive facilities 3.5. Waste disposal	Sec. 58.49 Laboratory operation areas. Sec. 58.51 Specimen and data storage facilities. **Subpart D – Equipment** Sec. 58.61 Equipment design. Sec. 58.63 Maintenance and calibration of equipment.
4. Apparatus, material, and reagents	**Subpart E – Testing Facilities Operation** Sec. 58.81 Standard operating procedures.
5. Test systems 5.1. Physical/chemical 5.2. Biological	Sec. 58.83 Reagents and solutions. Sec. 58.90 Animal care.
6. Test and reference items 6.1. Receipt, handling, sampling and storage 6.2. Characterisation	**Subpart F – Test and Control Articles** Sec. 58.105 Test and control article characterization. Sec. 58.107 Test and control article handling. Sec. 58.113 Mixtures of articles with carriers.
7. Standard operating procedures	**Subpart G – Protocol for and Conduct of a Nonclinical Laboratory Study** Sec. 58.120 Protocol. Sec. 58.130 Conduct of a nonclinical laboratory study.

(Continued)

Table 4.1 (*Continued*)

EU Directive 2004/10/EC	US 21 CFR Part 58
8. Performance of the study 8.1. Study plan 8.2. Content of the study plan 8.3. Conduct of the study	
9. Reporting of study results 9.1. General 9.2. Content of the final report **10. Storage and retention of records and materials**	**Subpart J – Records and Reports** Sec. 58.185 Reporting of nonclinical laboratory study results. Sec. 58.190 Storage and retrieval of records and data. Sec. 58.195 Retention of records.

understanding of the pharmacology of the drug, and address safety and quality concerns before engaging in trials with human subjects. The studies must provide detailed knowledge of the drug's pharmacodynamics and pharmacokinetics, in a combination of *in-vitro* and animal models, and permit adequate assessment of the toxicological properties of the drug. The drug substance and product must be characterised in terms of its chemical/physical characteristics, manufacture, control and stability. All safety studies must be conducted according to the principles of GLP.

4.10
Further Reading

- ICH Harmonised Guidance Documents
  www.ich.org.
- European Medicines Agency non-harmonised guidance documents
  http://www.emea.europa.eu/htms/human/humanguidelines/nonclinical.htm.
- FDA non-harmonised guidance documents
  www.fda.gov.
- GLP Directive 2004/10/EC
  http://www.europa.eu.int/eur-lex/index.html.
- US GLP Regulations, Code of Federal Regulations, Title 21, Part 58.
  www.fda.gov.

5
Clinical Trials

5.1
Chapter Introduction

This chapter examines the process of conducting clinical trials with human subjects. This is probably the most critical stage in the regulatory pathway, as the results establish the safety and efficacy of the drug in humans and ultimately determine whether or not a successful marketing authorisation application can be made. They also mark the stage where extensive engagement with the regulatory authorities in respect of the drug under development commences. There are considerable regulations governing the conduct of clinical trials, the underlying purposes of which are to ensure that the safety, welfare and rights of the subjects participating in the trials are protected, and that scientifically useful and valid data are collected.

5.2
Clinical Trials

The overall objective of clinical trials is to establish a drug therapy that is safe and effective in humans, to the extent that the risk–benefit relationship is acceptable. The ICH process has developed an internationally accepted definition of a clinical trial as:

> "Any investigation in human subjects intended to discover or verify the clinical, pharmacological and/or other pharmacodynamic effects of one or more investigational medicinal product(s), and/or to identify any adverse reactions to one or more investigational medicinal product(s) and/or to study absorption, distribution, metabolism and excretion of one or more investigational medicinal product(s) with the object of ascertaining its (their) safety and/or efficacy."

The process of conducting clinical trials for drug approval may be broken down into three consecutive phases of:

Phase I
Phase II
Phase III

Medical Product Regulatory Affairs. John J. Tobin and Gary Walsh
Copyright © 2008 WILEY-VCH Verlag GmbH & Co. KGaA, Weinheim
ISBN: 978-3-527-31877-3

Additional safety studies, termed Phase IV trials, may be conducted post-approval. Each phase may be distinguished in terms of its specific purpose and relative complexity. A drug may not proceed to the next trial phase until the results of the current phase have been analysed and reported, as the conclusions drawn will usually influence the study plan for the following phase. There are no absolute requirements as regards specific studies that must be conducted in each phase, as these may vary depending on the characteristics of the particular drug under investigation. Typically however, studies progress in the following manner.

5.2.1
Phase I Trials

The primary objective of a Phase I trial is to assess the safety of the drug in humans. Studies are normally conducted in healthy male volunteers, although specific categories of subject may be used in certain cases. For example, to avoid the risk of low blood pressure, subjects with mild hypertension would be more appropriate for the evaluation of antihypertensive drugs, while patients are likely to be used in the case of drugs that are expected to produce significant toxic effects (e.g. anti-cancer cytotoxic drugs). Remuneration may be offered for participation in the study. The number of subjects is normally between 10 and 100 people.

The study will commence with the administration of low doses, as judged from the non-clinical data. As the study progresses – and provided that there are no indications that it is unsafe to do so – the dosage levels may be increased past the anticipated therapeutic range. Subjects are closely monitored for changes in vital signs (blood pressure, heart rate, body temperature, etc.) and the emergence of any adverse side effects (nausea, drowsiness, pain, headache, irritability, hair loss, etc.).

Samples of blood and excreta are taken for laboratory analysis. It is expected that, at the end of this phase, you will have a preliminary estimate of the maximum dose that may be safely tolerated in humans, and also a basic profile of the drug's pharmacokinetic behaviour. Depending on the availability of appropriate analytical indicators, pharmacodynamic and indicative efficacy data may also be generated. The data acquired must be carefully analysed and assessed so that, based on the findings, appropriate Phase II trials can be planned.

Phase I trials are relatively short, and are usually completed within a year.

5.2.2
Phase II Trials

The primary objective of Phase II trials is to explore therapeutic efficacy in patients. The number of subjects may range from 50 to 500, and patients may be selected based on specific restrictive criteria; for example, those suffering only from the condition under study, disease stage, age, and so on. An important goal for this phase is to determine the dosage level and regimen for Phase III trials. Trials may commence with short range-finding studies in a small number of patients before more extensive studies are conducted. Again, subjects will be monitored for any undesirable side effects or changes in vital signs. However, the main thrust of the studies is to show that beneficial therapeutic effects may be obtained, and to determine the basic dose–response behaviour of the drug in humans. The studies may also involve identifying or validating convenient markers or endpoints that may be used as indicators of efficacy for later studies. In order to provide objective evidence of beneficial therapeutic effect, patient groups receiving the therapy are compared against reference or control groups that have received either an inactive placebo or the existing standard treatment.

Phase II trials may last from 1 to 2 years.

5.2.3
Phase III Trials

The purpose of Phase III trials is to confirm the safety and therapeutic efficacy of a drug in a broader range of subjects. This can involve studying groups that differ in terms of age, gender, disease stage, additional medical or physiological conditions (e.g. kidney or liver failure, pregnancy, breast-feeding), existing drug therapy, or ethnic background. Thus, the number of subjects required will be significantly greater than for Phase II trials and, depending on the specific drug, can easily approach or surpass 2000 people. In order to obtain the number and diversity of subjects, multi-centre – or indeed multi-country – studies are common at this stage. This is also the phase where trials involving extended exposure are normally conducted for drugs intended for administration over long periods. Thus, this phase may take from 3 to 5 years, or even longer, to complete. The Phase III studies provide a comprehensive understanding of the safety and efficacy of the drug, and on completion the drug developer should have identified and confirmed:

the type of patients/conditions that benefit from the therapy (indications for use);
the situations/conditions where the drug should not be prescribed (contrain-dications);
the appropriate dosage regimen; and
possible adverse side effects

A summary of the clinical trial process is shown in Figure 5.1. Data from clinical trials form the most critical part of the submission for marketing authorisation. In the case of drugs with multiple therapeutic uses, the developer may choose to submit an application for just one indication for use, while continuing with trials to support other indications. Parallel to the clinical trials, outstanding non-clinical studies will be continued. This may involve the optimi-sation of a drug formulation as dictated by clinical trial results, and the conducting of stability studies on the final drug product.

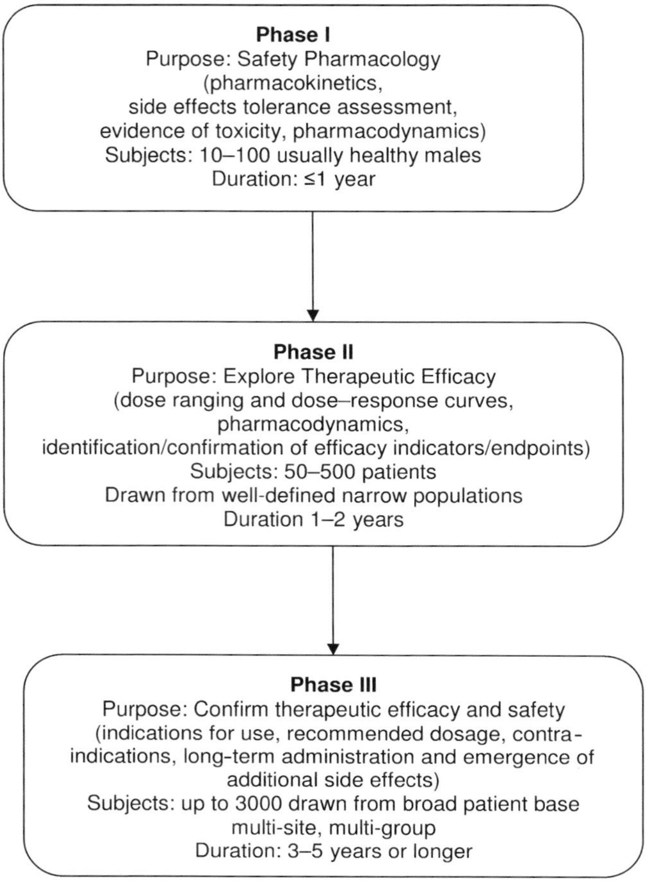

Phase I
Purpose: Safety Pharmacology
(pharmacokinetics,
side effects tolerance assessment,
evidence of toxicity, pharmacodynamics)
Subjects: 10–100 usually healthy males
Duration: ≤1 year

Phase II
Purpose: Explore Therapeutic Efficacy
(dose ranging and dose–response curves,
pharmacodynamics,
identification/confirmation of efficacy indicators/endpoints)
Subjects: 50–500 patients
Drawn from well-defined narrow populations
Duration 1–2 years

Phase III
Purpose: Confirm therapeutic efficacy and safety
(indications for use, recommended dosage, contraindications, long-term administration and emergence of
additional side effects)
Subjects: up to 3000 drawn from broad patient base
multi-site, multi-group
Duration: 3–5 years or longer

Figure 5.1 Summary of clinical trials conducted as a prelude to a marketing authorisation application.

5.3
Clinical Trial Design

A key element in planning and conducting clinical trials is to ensure that they have scientific validity and objectivity. This is particularly relevant with respect to Phase II and III studies, where it is desired to demonstrate a positive benefit to risk outcome. Responses to a drug among a patient population are rarely homogeneous and clear-cut. Thus, sound statistical principles must be applied in order to be able to distinguish significant effects from random events.

As a starting point, the objectives of each individual study should be clearly defined. Studies can be broadly divided into two types: (i) confirmatory studies which set out to prove or disprove predefined hypotheses on the basis of measuring defined variables (indicators or endpoints); and (ii) exploratory studies where the outcome in terms of indicators may not be clear in advance. Confirmatory trials can set out to show superiority of the treatment over a reference treatment/no treatment or equivalence (non-inferiority) to an existing established treatment.

The number of participants must be chosen so that the trial will have sufficient "statistical power", particularly in the case of confirmatory trials. As a general principle, the greater the number of data, the greater the confidence there is in demonstrating that a statistical difference between two groups exists, or not. However, numbers of participants will be constrained by cost considerations, the availability of suitable subjects and, above all, by the ethical principle that subjects should not be enrolled in clinical trial unless they add scientific value.

Clinical trials should be designed so as to minimise potential sources of bias. It is known that patients can demonstrate a positive response to treatments that they believe will benefit them, even if no pharmaceutical agent has been administered (the "placebo effect"). Similarly, investigators may be biased in their observations by an expectation of particular results. To avoid such bias, "blinded" trial designs are used.

In a *single-blind trial*, test subjects receive the drug, while the controls receive an apparently identical preparation that does not contain the active pharmaceutical ingredient under investigation. Subjects are not informed as to which treatment they have received. In a *double-blind trial* this information is withheld from both the subjects and the investigators that conduct the study. A further source of bias may be introduced, when selecting patients for control or test treatments. This can be avoided by using a randomised procedure, where once a subject meets the basic study criteria, they are enrolled and then assigned to control or test treatments according to a pre-set randomisation plan.

The most common clinical trial design for confirmatory trials is the *parallel group design*, in which subjects are randomised between a control group and one or more test groups. The results from the test groups are then compared to those of the control group. An ethical issue may arise with this type of design in situations where the treatment is very beneficial and thus it would be unfair to withhold it from a control group of patients. In such instances, it may be possible to use historical control data from studies that were previously conducted, but caution must be exercised to ensure that the data have relevance to the present day.

Crossover trial designs can be attractive in situations where the availability of suitable patients is limited. In a crossover trial, patients act as their own controls as they receive the placebo and test at different stages over the course of the study. In the simplest 2×2 crossover design, each subject receives each of two treatments in randomised order in two successive treatment periods (i.e. placebo-test-placebo-test, or test-placebo-test-placebo). These types of trial can yield statistically powerful data with small numbers, because aside from avoiding the requirement to have a separate control group, they eliminate the subject-to-subject variability that would be seen in a parallel group study. This type of study design is only suitable for drugs that are intended for the long-term treatment of chronic stable diseases, as opposed to drugs that can cure a condition. A washout interval between each treatment period is usually required to prevent drug carry over affecting the next stage of the study.

Table 5.1 A simple factorial trial design involving two drugs.

Treatment group	Treatment
1	Drug A + Drug B
2	Drug A + Placebo
3	Drug B + Placebo
4	Placebo + Placebo

Factorial trial designs are used to evaluate, simultaneously, two or more treatments through the use of varying combinations of the treatments. An example of a simple trial to evaluate drugs A and B is illustrated in Table 5.1. These trials are particularly relevant where multiple drug therapies are anticipated.

5.4
Good Clinical Practice

Clinical trials must be conducted in accordance with the principles of Good Clinical Practice (GCP). GCP is an international ethical and scientific quality standard for designing, conducting, recording and reporting trials that involve the participation of human subjects. Its purpose is twofold:

- To ensure that that the rights, safety and well-being of trial subjects are protected, in accordance with ethical principles that were developed by the World Medical Organisation and set out in the Declaration of Helsinki.
- To assure that the clinical trial data are credible.

A harmonised guidance document on GCP has been developed by the ICH (ICH E6). The core principles of GCP, as agreed in the guidance document, are outlined in Table 5.2.

These principles of GCP have been incorporated into regulations that govern the initiation and conduct of clinical trials. The process for undertaking clinical trials in the EU and the US shall now be examined.

5.5
Clinical Trials in the EU

Two directives were recently introduced to harmonise the conduct of clinical trials in the EU. These are:

- Directive 2001/20/EC on the approximation of the laws, regulations and administrative provisions of the Member States relating to the implementation of good clinical practice in the conduct of clinical trials on medicinal products for human use

Table 5.2 The core principles of Good Clinical Practice.

Helsinki Declaration Ethical Principles	Clinical trials shall be conducted in accordance with the Declaration of Helsinki on Ethical Principles for Medical Research Involving Human Subjects
Primacy of Subject's Rights	The rights, safety, and well-being of the trial subjects should prevail over the interests of science or society
Favourable Risk/benefit Balance	Prior assessment should indicate that the anticipated benefits would outweigh any foreseeable risks or inconveniences
Professional Medical Care (Investigator)	The medical care given to, and medical decisions made on behalf of, subjects should always be the responsibility of qualified medical personnel
Informed Consent	Freely given informed consent should be obtained from every subject
Subject Confidentiality	The confidentiality of records that could identify subjects should be protected to assure their privacy
Independent Ethics Committee Oversight	The trials must be accepted by an Independent Ethics Committee (IEC)/Institutional Review Board (IRB) before commencing the study
Supporting Data	Each proposed trial should be supported by existing non-clinical or clinical data
Scientifically Sound Protocol	Clinical trials should be scientifically sound, and conducted according to a clear, detailed protocol
Qualified Personnel	Individuals involved in conducting a trial should be qualified by education, training, and experience to perform his or her respective task(s)
Record Keeping	All clinical trial information should be recorded, handled, and stored in a way that allows its accurate reporting, interpretation and verification
Quality Assurance	Systems with procedures that assure the quality of every aspect of the trial should be implemented
GMP manufacture of investigational drug	Investigational products should be manufactured, handled, and stored in accordance with applicable good manufacturing practice (GMP)

- Directive 2005/28/EC, which lays down the principles and detailed guidelines for GCP as regards investigational medicinal products for human use, as well as the requirements for authorisation of the manufacturing or importation of such products

Prior to this, there was considerable variation across Member States as regards the requirements contained in national laws. Clinical trials are still regulated via national laws, but now these must contain the harmonised requirements established by the directives. The directives are quite general in nature, but are supported by detailed guidelines (ICH- and EU-specific) that may be accessed as part of "The Rules Governing Medicinal Products in the European Union, Volume 10 – Clinical Trials". The list of available ICH efficacy guidelines is shown in Figure 5.2. The process and players involved in conducting clinical trials are examined in the following paragraphs.

Clinical Safety
E1   The Extent of Population Exposure to Assess Clinical Safety for Drugs Intended for Long-Term Treatment of Non-Life Threatening Conditions
E2A   Clinical Safety Data Management: Definitions and Standards for Expedited Reporting
E2B(R3) Clinical Safety Data Management: Data Elements for Transmission of Individual Case Safety Reports
E2C(R1) Clinical Safety Data Management: Periodic Safety Update Reports for Marketed Drugs
E2D   Post-Approval Safety Data Management: Definitions and Standards for Expedited Reporting
E2E   Pharmacovigilance Planning
Clinical Study Reports
E3   Structure and Content of Clinical Study Reports
Dose-Response Studies
E4   Dose-Response Information to Support Drug Registration
Ethnic Factors
E5(R1)   Ethnic Factors in the Acceptability of Foreign Clinical Data
Good Clinical Practice
E6(R1)   Good Clinical Practice
Clinical Trials
E7   Studies in Support of Special Populations: Geriatrics
E8   General Consideration of Clinical Trials
E9   Statistical Principles for Clinical Trials
E10   Choice of Control Group and Related Issues in Clinical Trials
E11   Clinical Investigation of Medicinal Products in the Pediatric Population
Guidelines for Clinical Evaluation by Therapeutic Category
E12   Principles for Clinical Evaluation of New Antihypertensive Drugs
Clinical Evaluation
E14   The Clinical Evaluation of QT/QTc Interval Prolongation and Proarrhythmic Potential for Non-Antiarrhythmic Drugs
Pharmacogenomics
E15   Terminology in Pharmacogenomics

Figure 5.2 ICH efficacy guidelines.

5.5.1
The Sponsor

A *sponsor* initiates the process of conducting clinical trials. The sponsor may be an individual, a company, or an institution or organisation, but is most usually the drug development company or a contract research organisation that acts on its behalf. The sponsor is responsible for the initiation, management and/or financing of a clinical trial. The management responsibilities include making sure that the trials are conducted according to the regulations and GCP principles, and that the data are documented, analysed and reported according to established SOPs and appropriate quality control and quality assurance measures. The sponsor or his/her legal representative must be located in the EU.

5.5.2
The Investigator's Brochure

As a first step, the sponsor is required to prepare an *investigator's brochure* (IB). This must be an objective compilation of available clinical and non-clinical data on the

Table 5.3 Suggested headings for an Investigator's Brochure according to ICH guidelines.

1	**Table of Contents**
2	Summary
3	Introduction
4	Physical, Chemical, and Pharmaceutical Properties and Formulation
5	Non-clinical Studies
5.1	Non-clinical Pharmacology
5.2	Pharmacokinetics and Product Metabolism in Animals
5.3	Toxicology
6	Effects in Humans
6.1	Pharmacokinetics and Product Metabolism in Humans
6.2	Safety and Efficacy
6.3	Marketing Experience
7	Summary of Data and Guidance for the Investigator

investigational medicinal product that are relevant to the study of the product in human subjects. It should include a description of the physical, chemical and pharmaceutical properties of the drug substance and product, as well as all the relevant studies of its pharmacology, toxicology, pharmacokinetics and metabolism. At the outset, in the case of a new drug substance, only non-clinical information will be available, but clinical data or other relevant human experience should be included for later trials. Its purpose is to provide the investigators, and others involved in the trial, with sufficient information to facilitate their understanding of the rationale for, and their compliance with, many key features of the trial protocol, such as the dose, dose frequency/interval, methods of administration and safety monitoring procedures, as well as alerting them to the possible risks and adverse reactions that may be encountered. The IB should be validated and updated at least once a year. Suggested headings for an IB are shown in Table 5.3.

5.5.3
The Investigator

An *investigator* is responsible for the conduct of the trial at the trial site. The investigator must be a doctor or other appropriately qualified medical personnel (e.g. dentist), and should have adequate education, training and experience appropriate to the study. They manage the recruitment of subjects, administration of drugs and the monitoring and recording of responses. Their overriding responsibility is to guard the health and welfare of the subjects. Where multiple investigators are involved at a trial site, a lead or principal investigator must be designated.

5.5.4
The Trial Protocol

The sponsor with input from the investigator must prepare a *trial protocol*. This document should describe the objective(s), design, methodology (including subject selection criteria), statistical considerations and organisation of the trial. An example of headings for a trial protocol is shown in Figure 5.3. The protocol may be amended as necessary.

5.5.5
The Investigational Medicinal Product Dossier

An *Investigational Medicinal Product Dossier* (IMPD) is intended to be more comprehensive than an IB, in that it should contain summaries of available quality data in addition to the safety and efficacy information that constitutes the main part of the IB. In total, it should provide information on the chemistry, manufacture, control and stability of the medicinal product, together with the results of non-clinical and clinical studies. In order to avoid repetition, the IB can be cross-referenced for non-clinical and clinical results. Ideally, the IMPD should follow the same structure as that which will be used later for the marketing authorisation application. For products with existing marketing authorisations, the Summary of Product Characteristics may replace the IMPD to varying extents (see Chapter 6).

5.5.6
Informed Consent

All subjects enrolling on a trial must be informed in simple unbiased language of the purpose, risks, benefits, side effects and any other information that could be material to their decision to participate, including their right to withdraw at any time. Then, if agreeable, subjects – or their legal representatives in the case of minors or other persons unable to give informed consent – must sign and date an appropriate informed consent form. Financial inducements, other than as compensation, are prohibited in trials on minors or incapacitated adults. Trials involving such subjects should be designed so as to minimise pain, discomfort, or fear.

5.5.7
Manufacture of Investigational Medicinal Product

The manufacture of all investigational medicinal products (including placebo) that are intended for a clinical trial must be authorised and conducted according to Good Manufacturing Practice (GMP). This should be supervised and certified by a "Qualified Person". An import authorisation is required for any product from outside the EU, which should also be manufactured to GMP standards. (See Chapters 11 and 12 for information on GMP, manufacturing authorisations and Qualified Persons.)

Batches should be tailor-made for the proposed trial, and should be labelled with information as appropriate regarding the product, sponsor, investigator, subject and trial, together with the statement "For clinical trial use only". In addition, where the medicine will be used outside of the clinical site, the statement "Keep out of reach of children" should be added. There should be full accountability for the distribution, storage and fate of the drug product at both the manufacture and trial sites.

1.TITLE PAGE

Protocol Title: Protocol Code No: Protocol version:
Co-ordinating Investigator:(Multi-centre trial) / Principal Investigator:(Single centre trial)
Sponsors:
Other relevant personnel:
Study site(s);

2. TABLE OF CONTENTS

Please list with relevant pagination the contents of this application.

3. LIST OF ABBREVIATIONS

4. STUDY SYNOPSIS

Title
Investigational medicinal Product(s)Test and comparator:
Study Objectives:
Study Design:
Inclusion/Exclusion Criteria
Primary and Secondary Efficacy Endpoints:
Safety Endpoints:

5. BACKGROUND INFORMATION WITH RELEVANT BIBLIOGRAPHY AND RESULTS OF CLINICAL TRIALS RELEVANT TO THE APPLICATION.

Name and description of the investigational product(s). A summary of findings from non-clinical studies and from clinical trials that is relevant to the trial. Summary of the known and potential risks and benefits, if any, to human subjects. Description of and justification for the route of administration, dosage, dosage regimen, and treatment period(s) .A statement that the trial will be conducted in compliance with the protocol, GCP and the applicable regulatory requirement(s). Description of the population to be studied. References to literature and data that are relevant to the trial, and that provide background for the trial

6. OBJECTIVES OF THE TRIAL

Primary:
Secondary:

7. STUDY DESIGN

Appropriate study design should be selected to achieve the desired outcome. A description of the type/design of trial to be conducted (e.g., double-blind, placebo-controlled, parallel design) and a schematic diagram of trial design.
Start date
End of study date

8. PARTICIPANTS

Please describe planned study population. Choice of participants should be appropriate for the indication proposed. Subjects should not have enrolled in a clinical trial during the preceding 12 weeks. Please describe how special populations such as women of childbearing age, children and the elderly will be handled in this study. Please comment on the special need for close monitoring due to safety considerations.

9. INCLUSION CRITERIA / EXCLUSION CRITERIA

If the medicinal product (s) is currently licensed it is recommended that the current summary of product characteristics (Previously known as the data sheet) is consulted for appropriate contra indications, precautions and warnings.

10. INVESTIGATIONAL MEDICINAL PRODUCT

Test and comparator
Dosage forms and strengths:
Dose Schedule:
Route of Administration:
Duration of treatment:

11. CONCOMITANT MEDICATION:

Please outline any concomitant medications that are permitted for the duration of the trial. If the medicinal product (s) is currently licensed it is recommended that the current summary of product characteristics (Previously known as the data sheet) is consulted for information on potential drug interactions.

12. TREATMENT COMPLIANCE:

Please indicate how compliance with study treatment will be assessed and monitored.

Figure 5.3 A typical clinical trial protocol template (in this case as required by the Irish Medicines Board; www.imb.ie © Irish Medicines Board).

13. SAFETY

Please detail how safety data will be collected. Safety data should be collected according to the definitions of an adverse event and a serious adverse event as outlined in ICHE2A

14. DEFINITION OF ADVERSE AND SERIOUS ADVERSE EVENTS

15. PROCEDURES FOR MONITORING AND RECORDING ADVERSE EVENTS AND SERIOUS ADVERSE EVENTS

16. STATISTICAL ANALYSIS PLAN

We recommend that consideration be given to the statistical considerations involved in both the design and analysis of proposed research studies prior to commencing the study.

17. JUSTIFICATION OF POPULATION AND METHOD OF ESTIMATION:

The number of subjects planned to be enrolled, if more than one site the numbers of enrolled subjects projected for each trial site should be specified. Reason for choice of sample size; include calculations of the statistical power of the trial, the level of significance to be used and the clinical justification.

18. DEFINITION OF ITT AND PP POPULATION

Please provide a definition of the per protocol population and the intention to treat population

Per protocol population

Intention to treat population

19. CRITERIA AND PROCEDURE FOR DEALING WITH WITHDRAWALS FROM STUDY.

1. Criteria and procedure for withdrawal of subjects from the trial/ investigational product treatment
2. The nature and periodicity of the data to be collected for withdrawn subjects.
3. Replacement of withdrawn subjects.
4. The follow-up for subjects withdrawn from investigational product treatment/trial treatment. Should be decided in advance.

20. HANDLING OF DATA FROM SUBJECTS WITHDRAWN

The statistical handling of data from withdrawn patients should be defined a priori.

21. CONTROL OF BIAS

22. METHOD OF RANDOMISATION

Please describe the unit of randomisation, how the allocation schedule was devised and the method of execution of the randomisation process.

23. METHOD OF BLINDING

If applicable please outline the method of blinding proposed for this study. Please describe the similarity of treatment characteristics and evidence of successful masking among all involved in the study

24. EFFICACY ENDPOINTS:

These may be primary or secondary. Please indicate the minimum important differences. c.f. justification of population size. Please outline methods and timing for assessing, recording, and analysing of efficacy parameters.

25. SAFETY ENDPOINTS:

Methods and timing for assessing, recording, and analysing of safety parameters. Procedures for eliciting reports of and for recording and reporting adverse event and intercurrent illnesses.

26. CONSENT FORM AND PATIENT INFORMATION LEAFLET

Please provide an appropriate patient information leaflet providing a full explanation of the nature, purpose, procedures and risks associated with the study and consent form. Volunteers will sign the consent form a minimum of 6 days in advance of the study commencing. A separate consent form must be provided for any clinical trial involving testing of genetic material or sampling of genetic material for use at a later date .The patient information leaflet should contain a contact number where the principal investigator may be contacted

Figure 5.3 *(Continued)*

5.5.8

Competent Authority Clinical Trial Application

Before a clinical trial can commence, it is necessary to submit a *Clinical Trial Application* (CTA) to the Competent Authority of each Member State, where it is proposed to run a trial. As a prelude to this, the sponsor must obtain a unique EudraCT number from the centralised EudraCT database, which is used to track basic information on all trials taking place within the EU. The application form (11 pages) may also be accessed from the EudraCT server. The application form must be accompanied by the information related to the subjects (informed consent), the trial (trial protocol), the investigational medicinal product (IB and IMPD) or the Summary of Product Characteristics (if the drug already has a marketing authorisation), facilities and staff and financial arrangements. A checklist of the information that may be required with the application is shown in Figure 5.4. The specific document requirements vary among the member states.

The Competent Authorities are allowed a maximum of 60 days to review the data, or 90 days in the case of trials involving medicinal products for gene therapy, somatic cell therapy (including xenogeneic cell therapy), and all medicinal products containing genetically modified organisms. Such trials require written authorisation from the Competent Authority as distinct from other types of product where the authorities may just notify the sponsor of the acceptability (no-objection) of the application. If issues are raised, the sponsor may amend the application once, with a consequent extension to the allowed review period.

5.5.9
Independent Ethics Committee CTA

A favourable opinion from an Independent Ethics Committee is also required in order for a trial to proceed. The focus of such committees is to ensure that the rights, safety and welfare of human subjects involved in a trial are protected and, in doing so, to generate public confidence in the process. Ethics committee should consist of both healthcare professionals and lay members such as people with legal or religious backgrounds. They should have a formal legal status and operate according to defined procedures/standing orders. They may be organised on a national, regional or institutional basis.

Applications to the Ethics Committee should be made using the same application form as used for the Competent Authority. Most of the data submitted to the Competent Authority will also need to be included in the application to the Ethics Committee, although the IMPD is usually not required. Similar to the Competent Authority, the Ethics Committee is allowed 60 days to deliver an opinion (90 days for advanced therapy products), having given particular attention to the following:

- The relevance of the clinical trial and the trial design
- The anticipated benefits and risks justification
- The trial protocol
- The investigator's brochure
- The suitability of the investigator and supporting staff
- The quality of facilities
- The informed consent procedures and selection of subjects

1 General

1.1 Receipt of confirmation of EudraCT number

1.2 Covering letter

1.3 Application form

1.4 List of Competent Authorities within the Community to which the application has been submitted and details of decisions

1.5 Copy of ethics committee opinion in the MS concerned when available

1.6 Copy/summary of any scientific advice

1.7 If the applicant is not the sponsor, a letter of authorisation enabling the applicant to act on behalf of the sponsor

1.8 Language requirement; Does the CA require the application form to be completed in their national language, if not English?

2 Subject-related

2.1 Informed consent form

2.2 Subject information leaflet

2.3 Arrangements for recruitment of subjects

3 Protocol-related

3.1 Protocol with all current amendments

3.2 Summary of the protocol in the national language

3.3 Peer review of trial when available, not compulsory

3.4 Ethical assessment made by the principal/coordinating investigator

4 IMP-related

4.1 Investigator's brochure

4.2 Investigational Medicinal Product Dossier (IMPD)

4.3 Simplified IMPD for known products.

4.4 Summary of Product Characteristics (SmPC) (for products with marketing authorisation in the Community)

4.5 Outline of all active trials with the same IMP

4.6 If IMP manufactured in EU and if no marketing authorisation in EU:

4.6.1 – Copy of the manufacturing authorization referred to in Art. 13(1) of the Directive stating the scope of this authorization

4.7 If IMP not manufactured in EU and if no marketing authorisation in EU:

4.7.1 Certification of the QP that the manufacturing site works in compliance with GMP at least equivalent to EU GMP or that each production batch has undergone all relevant analyses, tests or checks necessary to confirm its quality

4.7.2 Certification of GMP status of active biological substance

4.7.3 – Copy of the importer's manufacturing authorization as referred to in Art. 13(1) of the Directive

4.8 Certificate of analysis for test product in exceptional cases:

4.8.1 – Where impurities are not justified by the specification or when unexpected impurities (not covered by specification) are detected

4.9 Viral safety studies when applicable.

4.10 Applicable authorisations to cover trials or products with special characteristics (if available) e.g. GMOs, radiopharmaceuticals

4.11 TSE Certificate when applicable

4.12 Language requirement; Does the CA require examples of the label in their national language, if not English?

5 Facilities and staff-related

5.1 Facilities for the trial

5.2 CV of the coordinating investigator in the MS concerned (for multicentre trials)

5.3 CV of each investigator responsible for the conduct of a trial in a site in the MS concerned (principal investigator)

5.4 Information about supporting staff

6 Finance-related

6.1 Provision for indemnity or compensation in the event of injury or death attributable to the clinical trial

6.2 Any insurance or indemnity to cover the liability of the sponsor or investigator

6.3 Compensations to investigators

6.4 Compensations to subjects

6.5 Agreement between the sponsor and the trial site

6.6 Agreement between the investigators and the trial sites

6.7 Certificate of agreement between sponsor and investigator when not in the protocol

Figure 5.4 A checklist of documentation to be provided with a Clinical Trial Application.

- Insurance cover for subjects, sponsor and investigators
- The financial arrangements for subjects and investigators.

In the case of multi-centre, multi-state trials a single opinion should be delivered from each Member State. Only when a positive response has been obtained from an Ethics Committee and a Competent Authority can a trial commence.

5.5.10
Amendments to Clinical Trials

During the course of its conduct, the sponsor may need to amend a clinical trial, as a consequence of the emergence of new information. If the amendments are deemed to be substantial, in that they may impact on the safety of trial subjects, or change the interpretation of the scientific documents in support of the conduct of the trial, or are otherwise significant, the sponsor shall notify the Competent Authority and the Ethics Committee, using a Trial Amendment Form. The Ethics Committee are permitted 35 days to approve the amendment.

5.5.11
Case Report Forms

All protocol-required data are recorded at the trial site on *Case Report Forms* (CRFs), which may be in either electronic or hard-copy format. Data for individual subjects are recorded on separate CRFs. The CRFs are used to transfer trial data to the sponsor for analysis and evaluation.

5.5.12
Adverse Event Reporting

It can be expected that, during the course of a clinical trial, some *adverse events* will be observed with subjects. These can range from the appearance of anticipated side effects to serious adverse events. The reporting requirements for an event will depend on an assessment of its seriousness, cause and predictability. A serious adverse event or reaction is defined as any untoward medical occurrence or effect that at any dose results in death, is life-threatening, requires hospitalisation or prolongation of existing hospitalisation, results in persistent or significant disability or incapacity, or in a congenital anomaly or birth defect. Irrespective of the cause, the investigator must immediately report all serious adverse events to the sponsor, unless they are specifically excluded from immediate reporting by the protocol.

If the events are classified as Suspected Unexpected Serious Adverse Reactions (SUSARs) that are fatal or life-threatening, the sponsor must then report them to the Ethics Committee and the Competent Authority within 7 days. Other SUSARs must be reported within 15 days. The sponsor must also inform all other investigators involved in the trial. The Competent Authority is required to enter the information into the EudraCT database, which ensures that the information is shared among the Competent Authorities of all the Member States.

5.5.13
Annual Safety Report

In addition to the expedited reporting described above, sponsors shall submit, once a year throughout the clinical trial, or on request, a safety report to the Competent Authority and the Ethics Committee. This should cover SUSARs, other serious adverse reactions, and an analysis of the subjects' safety during the course of the trial.

5.5.14
Monitoring of Trials

In accordance to GCP, the sponsor should appoint *clinical trial monitors*. These act as the main communication interface between the sponsor and the trial site, and should regularly visit the site to oversee that the trials are being conducted and correctly documented in accordance with the protocol and GCP. Reports should be supplied to the sponsor after each visit. It is also good practice for the sponsor to establish an auditing system for independently verifying that the activities in relation to the collection and processing of data at the trial site, and at related laboratories or sponsor's facilities, are conducted in accordance with applicable protocols, procedures, regulations, GCP and GLP.

The Competent Authorities are obliged to appoint Inspectors for checking that all activities associated with clinical trials are conducted in compliance with the regulations. The inspectors can inspect any sites concerned with the clinical trial, particularly the trial site (GCP), the manufacturing site of the investigational medicinal product (GMP), any laboratory used for analyses in the clinical trial (GLP), and/or the sponsor's premises.

5.5.15
End of Trial

The sponsor must inform the Competent Authority and the Ethics Committee of the end of a trial within 90 days of its completion. The sponsor will then prepare a *trial report*, where the data are analysed and assessed, with conclusions presented.

5.5.16
Trial Master File

The sponsor and investigator, as appropriate, must retain all essential documents relating to the trial for at least 5 years after its completion. Documents should be securely archived, with restricted access to maintain subject confidentiality. Collectively, these constitute the Trial Master File, and facilitate any audit and evaluation of the conduct of a clinical trial and the quality of the data produced. A summary of essential documents is shown in Figure 5.5.

5.6
Clinical Trials in The US

The relevant regulations governing the conduct of clinical trials in the US are shown in Table 5.4. As they also reflect the principles of GCP, they are quite similar in requirements to those of the EU. However, because they apply to a single jurisdiction, they are framed to provide more prescriptive detail than can be found in the equivalent EU directives. Similarly, they are supported by the ICH- and FDA-specific guidelines. As most of the practices are the same as discussed in the previous section, the chapter will now just examine some of the aspects that are unique to the US regulations.

5.6.1
Investigational New Drug Application (IND)

Requests for permission to conduct clinical trials with pharmaceuticals in the US are termed Investigational New Drug Applications (INDs). The applications are actually a request for an exemption to supply a drug without a marketing authorisation. A cover sheet (Form 1571) must accompany the application. This cover sheet should also be used with each subsequent communication with the FDA, with each form consecutively numbered, starting at 000 for the initial submission. A copy of the form is shown in Figure 5.6.

The content of the application is quite similar to EU requirements, although the structure and terminology is slightly different. The main headings are shown in Table 5.5. However, the FDA will now also accept IND application dossiers structured according to the CTD format (see Chapter 6 for the CTD format).

A Form 1572 is used for investigators to summarise their educational qualifications and experience, and to make a required formal declaration as to their commitment to conduct the study according to the protocol, GCP and the regulations. The sponsor should also collect financial disclosure information from the investigators at this stage, although formal declarations on Form 3455 are not required until the submission of a marketing authorisation application.

Once the FDA receive the initial submission, an IND reference number is assigned. The application is then passed on to the appropriate review centre; either the Center for Drug Evaluation and Research (CDER), or the Center for Biologics Evaluation and Research (CBER). Various experts will then review the submitted documents, the purpose being to ensure that the safety of subjects is not compromised and, in the case of Phase II and III studies, that the quality of study design is scientifically adequate. The FDA are allowed 30 days to complete the initial review, after which the study can commence, provided that it has been approved by an Institutional Review Board (IRB).

1. **PRE- TRIAL DOCUMENTS**
1.1 Investigator's Brochure (IB)
1.2 Signed protocol and amendments, if any and sample Case Report Form (CRF)
1.3 Information given to trial subject
    - Informed consent form
    - Any other written information
    - Advertisement for subject recruitment
1.4 Financial aspects of the trial
1.5 Insurance statement
1.6 Signed agreement between involved parties
1.7 Dated, documented approval / favourable opinion of Institutional Review Board / Independent Ethics Committee
1.8 Institutional Review Board / Independent Ethics Committee composition
1.9 Regulatory authority authorisation / approval / notification of protocol
1.10 Curriculum vitae and/or other relevant documents evidencing qualifications of investigators and sub-investigators
1.11 Normal values / ranges for medical / laboratory /technical procedures and/or tests included in the protocol
1.12 Medical / laboratory / technical procedures / tests
1.13 Sample of labels attached to investigational product containers
1.14 Instructions for handling of investigational products and trial-related materials
1.15 Shipping records for investigational product and trial-related materials
1.16 Certificates of Analysis of investigational product shipped
1.17 Decoding procedures for blinded trials
1.18 Master randomisation list
1.18 Pre-trial monitoring report
1.19 Trial initiation monitoring report

2. **DOCUMENTS GENERATED DURING THE COURSE OF THE TRIAL**
2.1 Investigator's Brochure updates
2.2 Revisions to protocol, CRF, informed consent form, etc.
2.3 Dated, documented approval / favourable opinion of Institutional Review Board / Independent Ethics Committee of amendments
2.4 Regulatory Authority authorisations /approvals / notifications where required
2.5 Curriculum vitae for new investigators and/or sub-investigators
2.6 Updates to normal values / ranges for medical / laboratory /technical procedures and/or tests included in the protocol
2.7 Updates of medical / laboratory / technical procedures / tests
2.8 Documentation of investigational products and trial-related materials shipment
2.9 Certificates of analysis for new batches of investigational products
2.10 Monitoring visit reports
2.11 Relevant communications other than site visits
2.12 Signed informed consent forms
2.13 Source documents
2.14 Signed, dated and completed Case Report Forms (CRF)
2.15 Documentation of CRF corrections
2.16 Notification by originating investigator to sponsor of serious adverse events and related reports
2.17 Notification by sponsor and/or investigator, where applicable to regulatory authorities and IRBs/IECs of unexpected serious adverse drug reactions and other safety information
2.18 Notification by sponsor to investigators of safety information
2.19 Interm or annual reports to IRB/IEC and authorities
2.20 Subject screening log
2.21 Subject identification code list
2.22 Subject enrolment log
2.23 Investigational products accountability at the site
2.24 Signature sheet
2.25 Record of retained body fluids / tissue samples

Figure 5.5 Essential documentation for a Trial Master File.

3.   POST-TRIAL DOCUMENTS
3.1   Investigational products accountability at the site
3.2   Documentation of investigational product destruction
3.3   Completed subject identification code list
3.4   Audit certificate (if available)
3.5   Final trial close-out monitoring report
3.6   Treatment allocation and decoding documentation
3.7   Final report by investigator to IRB/IEC where required, and where applicable, to the regulatory authorities
3.8   Clinical Study Report

Figure 5.5 *(Continued)*

5.6.2
Institutional Review Board

Under US regulations, each institution conducting research with human subjects must have its own IRB. These perform the same functions as the Independent Ethics Committees in Europe, and should contain at least five members, one of which should be independent of the institution. There should be available a mixture of scientific and non-scientific expertise capable of assessing research proposals from legal, ethical and scientific perspectives. The IRB must grant written authorisation to the investigator before a study can commence. They are also responsible for on-going reviews of research, and must report to the FDA:

- any unanticipated problems that place subjects at risk;
- any serious non-compliance with the regulations; or
- any decisions to suspend IRB approval.

Table 5.4 Code of Federal Regulations Parts relevant to Clinical Trials.

21 CFR Part 312	Investigational New Drug (IND)
	Sec. 312.6 Labeling of an investigational new drug
	Sec. 312.20 Requirement for an IND
	Sec. 312.21 Phases of an investigation
	Sec. 312.22 General principles of the IND submission
	Sec. 312.23 IND content and format
	Sec. 312.30 Protocol amendments
	Sec. 312.31 Information amendments
	Sec. 312.32 IND safety reports
	Sec. 312.33 Annual reports
	Sec. 312.41 Comment and advice on an IND
	Sec. 312.42 Clinical holds and requests for modification
	Sec. 312.47 Meetings
	Sec. 312.50 General responsibilities of sponsors
21 CFR Part 50	Protection of Human Subjects
21 CFR Part 54	Financial Disclosure by Clinical Investigators
21 CFR Part 56	Institutional Review Boards (IRBs)

DEPARTMENT OF HEALTH AND HUMAN SERVICES FOOD AND DRUG ADMINISTRATION	*Form Approved:* OMB No. 0910-0014. *Expiration Date: May 31, 2009* *See OMB Statement on Reverse.*
INVESTIGATIONAL NEW DRUG APPLICATION (IND) *(TITLE 21, CODE OF FEDERAL REGULATIONS (CFR) PART 312)*	**NOTE:** No drug may be shipped or clinical investigation begun until an IND for that investigation is in effect (21 CFR 312.40).

1. NAME OF SPONSOR	2. DATE OF SUBMISSION

3. ADDRESS *(Number, Street, City, State and Zip Code)*	4. TELEPHONE NUMBER *(Include Area Code)*

5. NAME(S) OF DRUG *(Include all available names: Trade, Generic, Chemical, Code)*	6. IND NUMBER *(If previously assigned)*

7. INDICATION(S) *(Covered by this submission)*

8. PHASE(S) OF CLINICAL INVESTIGATION TO BE CONDUCTED:
☐ PHASE 1 ☐ PHASE 2 ☐ PHASE 3 ☐ OTHER_____
(Specify)

9. LIST NUMBERS OF ALL INVESTIGATIONAL NEW DRUG APPLICATIONS (21 CFR Part 312), NEW DRUG OR ANTIBIOTIC APPLICATIONS *(21 CFR Part 314)*, DRUG MASTER FILES *(21 CFR Part 314.420)*, AND PRODUCT LICENSE APPLICATIONS (21 CFR Part 601) REFERRED TO IN THIS APPLICATION.

10. **IND submission should be consecutively numbered. The initial IND should be numbered "Serial number: 0000." The next submission (e.g., amendment, report, or correspondence) should be numbered "Serial Number: 0001." Subsequent submissions should be numbered consecutively in the order in which they are submitted.**	SERIAL NUMBER ___ ___ ___ ___

11. THIS SUBMISSION CONTAINS THE FOLLOWING: *(Check all that apply)*
☐ INITIAL INVESTIGATIONAL NEW DRUG APPLICATION (IND)   ☐ RESPONSE TO CLINICAL HOLD

PROTOCOL AMENDMENT(S):	INFORMATION AMENDMENT(S):	IND SAFETY REPORT(S):
☐ NEW PROTOCOL	☐ CHEMISTRY/MICROBIOLOGY	☐ INITIAL WRITTEN REPORT
☐ CHANGE IN PROTOCOL	☐ PHARMACOLOGY/TOXICOLOGY	☐ FOLLOW-UP TO A WRITTEN REPORT
☐ NEW INVESTIGATOR	☐ CLINICAL	

☐ RESPONSE TO FDA REQUEST FOR INFORMATION   ☐ ANNUAL REPORT   ☐ GENERAL CORRESPONDENCE

☐ REQUEST FOR REINSTATEMENT OF IND THAT IS WITHDRAWN,   ☐ OTHER _____
INACTIVATED, TERMINATED OR DISCONTINUED   *(Specify)*

CHECK ONLY IF APPLICABLE

JUSTIFICATION STATEMENT MUST BE SUBMITTED WITH APPLICATION FOR ANY CHECKED BELOW. REFER TO THE CITED CFR SECTION FOR FURTHER INFORMATION.

☐ TREATMENT IND 21 CFR 312.35(b)   ☐ TREATMENT PROTOCOL 21 CFR 312.35(a)   ☐ CHARGE REQUEST/NOTIFICATION 21 CFR 312.7(d)

FOR FDA USE ONLY

CDR/DBIND/DGD RECEIPT STAMP	DDR RECEIPT STAMP	DIVISION ASSIGNMENT:
		IND NUMBER ASSIGNED:

FORM FDA 1571 (4/06)   PREVIOUS EDITION IS OBSOLETE.   PAGE 1 OF 2

PSC Graphics: (301) 443-1090   EF

Figure 5.6 Form FDA 1571, Investigational New Drug application (IND).

CONTENTS OF APPLICATION

12. This application contains the following items: *(Check all that apply)*

☐ 1. Form FDA 1571 *[21 CFR 312.23(a)(1)]*
☐ 2. Table of Contents *[21 CFR 312.23(a)(2)]*
☐ 3. Introductory statement *[21 CFR 312.23(a)(3)]*
☐ 4. General Investigational plan *[21 CFR 312.23(a)(3)]*
☐ 5. Investigator's brochure *[21 CFR 312.23(a)(5)]*
☐ 6. Protocol(s) *[21 CFR 312.23(a)(6)]*
   ☐ a. Study protocol(s) *[21 CFR 312.23(a)(6)]*
   ☐ b. Investigator data *[21 CFR 312.23(a)(6)(iii)(b)]* or completed Form(s) FDA 1572
   ☐ c. Facilities data *[21 CFR 312.23(a)(6)(iii)(b)]* or completed Form(s) FDA 1572
   ☐ d. Institutional Review Board data *[21 CFR 312.23(a)(6)(iii)(b)]* or completed Form(s) FDA 1572
☐ 7. Chemistry, manufacturing, and control data *[21 CFR 312.23(a)(7)]*
   ☐ Environmental assessment or claim for exclusion *[21 CFR 312.23(a)(7)(iv)(e)]*
☐ 8. Pharmacology and toxicology data *[21 CFR 312.23(a)(8)]*
☐ 9. Previous human experience *[21 CFR 312.23(a)(9)]*
☐ 10. Additional information *[21 CFR 312.23(a)(10)]*

13. IS ANY PART OF THE CLINICAL STUDY TO BE CONDUCTED BY A CONTRACT RESEARCH ORGANIZATION? ☐ YES ☐ NO

IF YES, WILL ANY SPONSOR OBLIGATIONS BE TRANSFERRED TO THE CONTRACT RESEARCH ORGANIZATION? ☐ YES ☐ NO

IF YES, ATTACH A STATEMENT CONTAINING THE NAME AND ADDRESS OF THE CONTRACT RESEARCH ORGANIZATION, IDENTIFICATION OF THE CLINICAL STUDY, AND A LISTING OF THE OBLIGATIONS TRANSFERRED.

14. NAME AND TITLE OF THE PERSON RESPONSIBLE FOR MONITORING THE CONDUCT AND PROGRESS OF THE CLINICAL INVESTIGATIONS

15. NAME(S) AND TITLE(S) OF THE PERSON(S) RESPONSIBLE FOR REVIEW AND EVALUATION OF INFORMATION RELEVANT TO THE SAFETY OF THE DRUG

I agree not to begin clinical investigations until 30 days after FDA's receipt of the IND unless I receive earlier notification by FDA that the studies may begin. I also agree not to begin or continue clinical investigations covered by the IND if those studies are placed on clinical hold. I agree that an Institutional Review Board (IRB) that complies with the requirements set fourth in 21 CFR Part 56 will be responsible for initial and continuing review and approval of each of the studies in the proposed clinical investigation. I agree to conduct the investigation in accordance with all other applicable regulatory requirements.

16. NAME OF SPONSOR OR SPONSOR'S AUTHORIZED REPRESENTATIVE

17. SIGNATURE OF SPONSOR OR SPONSOR'S AUTHORIZED REPRESENTATIVE   **Sign**

18. ADDRESS *(Number, Street, City, State and Zip Code)*

19. TELEPHONE NUMBER *(Include Area Code)*

20. DATE

(**WARNING**: A willfully false statement is a criminal offense. U.S.C. Title 18, Sec. 1001.)

Public reporting burden for this collection of information is estimated to average 100 hours per response, including the time for reviewing instructions, searching existing data sources, gathering and maintaining the data needed, and completing reviewing the collection of information. Send comments regarding this burden estimate or any other aspect of this collection of information, including suggestions for reducing this burden to:

Department of Health and Human Services
Food and Drug Administration
Center for Drug Evaluation and Research (HFD-143)
Central Document Room
5901-B Ammendale Road
Beltsville, MD 207052-1266

Department of Health and Human Services
Food and Drug Administration
Center for Biologics Evaluation and Research (HFM-99)
1401 Rockville Pike
Rockville, MD 20852-1448
Please DO NOT RETURN this application to this address.

"An agency may not conduct or sponsor, and a person is not required to respond to, a collection of information unless it displays a currently valid OMB control number."

FORM FDA 1571 (4/06)   PAGE 2 OF 2

Figure 5.6 *(Continued)*

Table 5.5 IND headings.

1	Cover sheet (Form FDA-1571)
2	A table of contents
3	Introductory statement
4	General investigational plan
5	Investigator's brochure
6	Protocols
7	Chemistry, manufacturing, and control information
8	Pharmacology and toxicology information
9	Previous human experience with the investigational drug
10	Additional information

5.6.3
Communication with the FDA

Once the initial IND application has been assigned a reference number, the sponsor can keep submitting information to the FDA file to enable the clinical development to move through the various trial phases.

Protocol amendments are used to inform the FDA of items such as changes to existing protocols, new protocols (for Phase II or III studies) or new investigators (trial sites). The changes can be implemented once they have been approved by the relevant IRB. However, the FDA can place a trial on "clinical hold" at any stage, if it is concerned over aspects such as safety risks to subjects, investigator qualifications, quality of the investigator's brochure, insufficient IND information to enable risk assessment, or deficiencies in design to achieve the objectives of the Phase II or III studies. Alternatively, the FDA may offer advice – or request clarification – if their concerns are of a less serious nature. The sponsor can also meet with the FDA before submitting the IND (pre-IND meeting) and, more critically, at the end of the Phase II studies. Here, he/she should seek to establish that the Phase III study designs proposed to confirm efficacy, will satisfy FDA expectations, when it comes to the point of submitting an application for marketing approval.

Information amendments are used to inform the FDA of new pharmacology or toxicology data, final non-clinical study reports, new chemistry-manufacturing-control information, or notice of the discontinuation of a clinical study.

The sponsor must also submit an *annual report* on the progress of the trial, within 60 days of the anniversary of the IND going into effect. This should provide brief summaries of the status of each study, the clinical and non-clinical information gathered during the previous year (including adverse reactions and safety reports), the general investigation plan for the coming year and, if applicable, changes to the investigator's brochure or Phase I protocol, or marketing authorisations/developments in other countries.

The sponsor is obliged to submit safety reports for SUSAR events to the same timelines as apply in Europe. They must also report within 15 days any findings from animal studies that would suggest an increased risk to human subjects, such as carcinogenicity or mutagenicity.

5.6.4
Labelling of Investigational Drugs

Drugs intended for use in clinical trials should bear a label containing the following statement:

- "Caution: New Drug–Limited by Federal (or United States) law to investigational use".

Drugs shipped for non-clinical trials should be labelled with one of the following statements:

- "CAUTION: Contains a new drug for investigational use only in laboratory research animals, or for tests in vitro. Not for use in humans"; or
- "CAUTION: Contains a biological product for investigational *in-vitro* diagnostic tests only".

5.7
Chapter Review

This chapter described the purpose and design of the different types of study conducted during the clinical phase of drug development. It then proceeded to examine the principles of Good Clinical Practice, and outlined how these are incorporated into the regulations governing the conduct of clinical trials in the EU. Finally, the chapter examined some of the distinctive aspects of the regulations that govern clinical trials in the US.

5.8
Further Reading

- ICH guidelines   Good Clinical Practice, ICH E6.
  www.ich.org.
- European Directives & Guidance on Clinical Trials
  Directives on Good Clinical Practice 2001/20/EC & 2005/28/EC.
  The Rules Governing Medicinal Products in the European Union, Volume 10, Clinical Trials.
  http://ec.europa.eu/enterprise/pharmaceuticals/index_en.htm.
- US IND Regulations   Code of Federal Register, Title 21, Part 312.
  www.fda.gov.

6
Marketing Authorisation

6.1
Chapter Introduction

This chapter looks at the process of obtaining a marketing authorisation, the final step in bringing a drug to the marketplace. This requires that the data collected during the non-clinical and clinical studies be organised into a comprehensive dossier for submission to the regulatory authorities. The authorities subject the dossier to a thorough review to ensure that there is satisfactory evidence as to the quality, safety and efficacy of the drug. They can then grant a marketing authorisation with confidence that the drug can contribute to public health, without undue risk. It is only then that commercial sales of the product can commence and the developer can start to recoup some of the expenditure required to bring it this far.

6.2
The Application Dossier

In order to obtain a marketing authorisation, the applicant must submit a dossier to the regulatory authorities containing extensive technical data on the quality, safety and efficacy of the drug, together with administrative and other documents such as application forms, declarations, labels and leaflets. The volume of information that must be assembled is quite large, and can easily run up to 120 000 pages of hard copy. In order to facilitate the review process, different regulatory authorities had previously specified their own format for how the dossier should be assembled. This meant that, although all regulatory authorities required essentially the same technical information, applicants had to expend considerable time and effort presenting the information in the format specified by each authority. However, through the International Conference on Harmonisation (ICH) process a common format that is acceptable to EU, US and Japanese regulatory authorities has been agreed. This is known as the Common Technical Document (CTD), and was recommended for adoption by the regulatory authorities in November 2000.

Medical Product Regulatory Affairs. John J. Tobin and Gary Walsh
Copyright © 2008 WILEY-VCH Verlag GmbH & Co. KGaA, Weinheim
ISBN: 978-3-527-31877-3

6.3
CTD

The CTD is organised into five modules. A schematic representation of the structure and hierarchy is shown in Figure 6.1. Module 1 is designed to contain region-specific information such as application forms and other administrative provisions that may apply. As such, it is not harmonised and is not considered part of the CTD. The other four modules present the technical data in a harmonised format. Module 2 should contain critical overview assessments of the quality, non-clinical and clinical data, together with summaries of the non-clinical and clinical data. The objective of this section is to provide reviewers with an introduction to the submission, and to orient

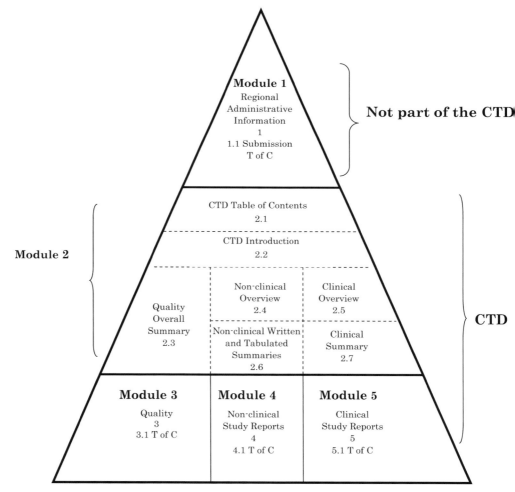

Figure 6.1 Diagrammatic representation of the organization of the Common Technical Document (CTD) (from ICH M4 Guide).

Table 6.1 Relationship between CTD and previous EU dossier structure.

CTD Structure		Previous EU Dossier Structure	
Module No	**Topic**	**Topic**	**Part No**
1	Regional Administrative Information	Summary of the Dossier	I
		Administrative	I a
		Summary of Product Characteristics	I b
2	Overviews/Summaries of Quality Non-Clinical and Clinical	Expert Reports	I c
3	Quality	Chemical, Pharmaceutical and Biological Testing of Medicinal Products	II
4	Non-Clinical Study Reports	Toxicological and Pharmacological Tests	III
5	Clinical Study Reports	Clinical Documentation	IV

them as to how the applicant believes the data support the granting of a marketing authorisation. Modules 3, 4 and 5 provide the detailed quality, non-clinical and clinical data, to which the reviewer may then refer.

The CTD format has been incorporated into the EU regulations and is set out in the revised version of Annex I to directive 2001/82/EC, which was published as Commission Directive 2003/63/EC. It is also supported by detailed guidance in the Rules Governing Medicinal Products in the European Union, Volume 2B - Presentation and content of the dossier - CTD2001 edition. The relationship between the CTD format and the previous four-section format that was used in the EU is shown in Table 6.1.

The Food and Drug Administration (FDA) also accept submissions prepared according to the CTD format. However, they have not directly incorporated the CTD format into US regulatory requirements, which are detailed in Title 21 CFR, Part 514.50 in the case of a New Drug Application (NDA), or Title 21 CFR Part 601.2 (a) in the case of a Biologics License Application (BLA). Rather, they have introduced the CTD format via "Guidance for Industry" documents. The correlation between the requirements in 21 CFR Part 514.50 and the CTD format is shown in Table 6.2.

Each module should be comprised of individual documents as the fundamental unit. Each document should have its own internal page numbering. If using a hard-copy submission, the documents should be bound together in volumes, separated by tabs. However, hard-copy submissions are being increasingly replaced by electronic submissions, as it makes it easier for the applicant and the regulatory authorities to handle. Again, through the ICH process specifications have been developed for the submission of an electronic Common Technical Document (eCTD). These basically address the directory structures and file formats that should be used to enable

Table 6.2 Correlation between the requirements of 21 CFR Part 314.50 and the CTD format.

21 CFR Part 314.50 Requirements	CTD
(a) Application form	1
(b) Index	2
(c) Summary	
(c)(2)(i) Annotated draft labelling	1
(c)(2)(ii) Pharmacologic class	2
(c)(2)(iii) Marketing history	1
(c)(2)(iv) CMC Summary	2
(c)(2)(v) Non-clinical pharmacology and toxicology summary	2
(c)(2)(vi) Human pharmacokinetics and bioavailability summary	2
(c)(2)(vii) Microbiology summary (Anti-infectives only)	2
(c)(2)(viii) Clinical data and statistics summary	2
(c)(2)(ix) Concluding Discussion	2
(d) Technical Sections	
(d)(1) Chemistry, manufacturing, and controls section	3
(d)(2) Non-clinical pharmacology and toxicology section	4
(d)(3) Human pharmacokinetics and bioavailability section	5
(d)(4) Microbiology section (anti-infectives only)	5
(d)(5) Clinical data section	5
(d)(6) Statistical section	5
(d)(7) Paediatric use section	5
(e) Samples of labels and drug (when requested)	1
(f) Case report forms and tabulations	5
(g) Other	
(h) Patent information	1
(i) Patent certification	1
(j) Claimed exclusivity	1
(k) Financial certification or disclosure statement	1

regulatory authorities to handle electronic submissions without encountering software issues. As such, the specifications have been developed to facilitate reviewers who may wish to:

- Copy and paste
- View and print documents
- Annotate documentation
- Export information to databases
- Search within and across applications
- Navigate throughout the eCTD and its subsequent amendments/variations.

Documents should be provided, where possible, as individual Portable Document Format (PDF) files, while Extensible Markup Language (XML) must be used to provide a user interface that enables navigation and viewing via a standard web browser. This offers the potential for an applicant to make a compete submission on just two or three CD-ROMs, as opposed to having to provide multiple copies if using paper submissions. At present, the FDA can accept single copy complete paperless submissions from applicants who satisfy their electronic signatures requirements, whereas the EU still require a complete paper copy as the official "master" copy. Applicants are advised to check with the authorities before making an electronic submission.

It is worthwhile to note that, in the US, where the applicant will just be dealing with a single authority, there is no need to re-submit data that was previously submitted as part of an IND application to conduct clinical trials. Instead, the applicant can cross-reference the IND file. This does not apply in Europe, because clinical trial applications will have been submitted to individual Competent Authorities, whereas marketing authorisation applications are usually submitted either centrally to the European Medicines Agency (EMEA) or collectively to a number of Competent Authorities. Thus, the files need to be self-supporting.

6.3.1
Module Structure

The ICH have issued guidelines on how to organise the data in the modules. These guidance documents, which are shown in Table 6.3, deal primarily with how the information should be organised, whereas advice on how to generate the data is provided in the quality, safety and efficacy guidelines that were discussed in the previous chapters. We shall now look at the structure of the modules in a little more detail, starting with the basic data modules.

6.3.2
Module 3: Quality

The quality module must contain information on the identity, characteristics, manufacturing methods, control, packaging and stability of both the drug substance and final drug product. The standard headings used to present this information are shown in Figure 6.2.

Table 6.3 ICH CTD guidance documents.

Document	Topics covered
M4(R3)	Common Technical Document Superstructure & document granularity
Quality – M4Q(R1)	Module 3: Quality + Module 2 Quality Overall Summary
Safety – M4S(R2)	Module 4: Non-clinical + Module 2 Non-Clinical Overview and Summary
Efficacy – M4E(R1)	Module 5: Clinical + Module 2 Clinical Overview and Summary
eCTD Specification	ICH Electronic Common Technical Document Specification, Version 3.2

3.1. Table of Contents of Module 3

3.2. Body of Data
   3.2.S Drug Substance
      3.2.S.1 General Information
         3.2.S.1.1 Nomenclature
         3.2.S.1.2 Structure
         3.2.S.1.3 General Properties
      3.2.S.2 Manufacture
         3.2.S.2.1Manufacturer(s)
         3.2.S.2.2Description of Manufacturing Process and Process Controls
         3.2.S.2.3 Control of Materials
         3.2.S.2.4 Controls of Critical Steps and Intermediates
         3.2.S.2.5 Process Validation and/or Evaluation
         3.2.S.2.6 Manufacturing Process Development
      3.2.S.3 Characterisation
         3.2.S.3.1 Elucidation of Structure and other Characteristics
         3.2.S.3.2 Impurities
      3.2.S.4 Control of Drug Substance
         3.2.S.4.1 Specification
         3.2.S.4.2 Analytical Procedures
         3.2.S.4.3 Validation of Analytical Procedures
         3.2.S.4.4 Batch Analyses
         3.2.S.4.5 Justification of Specification
      3.2.S.5Reference Standards or Materials
      3.2.S.6Container Closure System
      3.2.S.7Stability
         3.2.S.7.1Stability Summary and Conclusions
         3.2.S.7.2Post-approval Stability Protocol and Stability Commitment
         3.2.S.7.3Stability Data

   3.2.P Drug Product
      3.2.P.1 Description and Composition of the Drug Product
      3.2.P.2 Pharmaceutical Development
         3.2.P.2.1 Components of the Drug Product
            3.2.P.2.1.1 Drug Substance
            3.2.P.2.1.2 Excipients
         3.2.P.2.2 Drug Product
            3.2.P.2.2.1 Formulation Development
            3.2.P.2.2.2 Overages
            3.2.P.2.2.3 Physicochemical and Biological Properties
         3.2.P.2.3 Manufacturing Process Development
         3.2.P.2.4 Container Closure System
         3.2.P.2.5 Microbiological Attributes
         3.2.P.2.6 Compatibility
      3.2.P.3 Manufacture
         3.2.P.3.1 Manufacturer(s)
         3.2.P.3.2 Batch Formula
         3.2.P.3.3 Description of Manufacturing Process and Process Controls
         3.2.P.3.4 Controls of Critical Steps and Intermediates
         3.2.P.3.5 Process Validation and/or Evaluation
      3.2.P.4 Control of Excipients
         3.2.P.4.1 Specifications
         3.2.P.4.2 Analytical Procedures
         3.2.P.4.3 Validation of Analytical Procedures
         3.2.P.4.4 Justification of Specifications
         3.2.P.4.5 Excipients of Human or Animal Origin
         3.2.P.4.6 Novel Excipients
      3.2.P.5 Control of Drug Product
         3.2.P.5.1 Specification(s)

Figure 6.2 The headings structure of Module 3.

3.2.P.5.2 Analytical Procedures
3.2.P.5.3 Validation of Analytical Procedure(s)
3.2.P.5.4 Batch Analyses
3.2.P.5.5 Characterisation of Impurities
3.2.P.5.6 Justification of Specification(s)
3.2.P.6 Reference Standards or Materials
3.2.P.7 Container Closure System
3.2.P.8 Stability
3.2.P.8.1 Stability Summary and Conclusion
3.2.P.8.2 Post-approval Stability Protocol and Stability Commitment
3.2.P.8.3 Stability Data

3.2.A Appendices
3.2.A.1 Facilities and Equipment
3.2.A.2 Adventitious Agents Safety Evaluation
3.2.A.3 Excipients

3.2.R Regional Information
Executed Batch Records (USA only)
Method Validation Package (USA only)
Comparability Protocols (USA only)
Process Validation Scheme for the Drug Product (EU only)
Medical Device (EU only)

3.3. Literature References

Figure 6.2 (*Continued*)

- *Identity:* The drug substance should be identified as appropriate by its:

  o Recommended International Non-proprietary Name (INN);
  o Other non-proprietary names, such as United States Adopted Name (USAN), Japanese Accepted Name (JAN); British Approved Name (BAN);
  o Compendial name;
  o Chemical name;
  o Company or laboratory code (Proprietary name); and
  o Chemical Abstracts Service (CAS) registry number.

- *Characterisation:* Typical data that may be provided to describe the physical and chemical characteristics of the drug substance are shown in Figure 6.3. The description of the drug product should include the recipe detailing all the other ingredients (excipients) that are used to formulate the final product. It is vital to provide adequate data on the dissolution behaviour of the drug product, as the reviewer may wish to tie this in with pharmacokinetic or bioavailability data submitted in the other modules.

- *Manufacturing process:* The descriptions of the manufacturing steps for the drug substance and product should include process flow diagrams and discussions of critical scale-up steps and process development history and process validation activities, together with assessment of the equivalence or differences in batches used for various studies.

- *Control:* All control points starting with the basic raw materials right through to the finished product must be identified. Descriptions of the specifications, test methods, reference standards, and methods validation data should be included.

- Structural information: chemical formula and stereochemistry in the case of a New Chemical Entity (NCE) or amino acid sequence and glycosylation sites in the case of a biotech product

- Chemical and physical properties: appearance, colour, odour, molecular weight, melting point, refractive index, ionisation constants, partition coefficients, solubility under different conditions of pH or solvent

- Spectral properties: infrared, UV- visible, X-ray, NMR, etc.

- Chromatographic properties: TLC, HPLC, GC-MS

- Electrophoretic properties: Isoelectric point, SDS–PAGE profile

Figure 6.3 Typical data that may be submitted to characterise the drug substance.

- *Packaging:* Similarly, information on the closure/packaging systems must be provided in terms of material specification, suitability/compatibility with the pharmaceutical product, dimensional specifications, water impermeability, and so on. Defence against microbial contamination should be discussed in the context of either packaging of sterile product or use of preservatives as appropriate.

- *Stability:* Stability protocols and data should be submitted, along with the commitment and protocols to continue stability testing post-approval.

- *Appendices:* This section is most likely to contain additional data associated with biological-based products. It should contain information as regards the facilities and equipment used for the manufacture of biotech products. Assessment of the risk of contamination from adventitious agents such as transmissible spongiform encephalopathy agents (TSEs), bacteria, mycoplasma, fungi or viruses should also be provided. Additional information on novel excipients that have not been used before should also be included in this section.

- *Regional:* This section is used to include region-specific quality information. In a US submission, this should contain samples of executed batch records, the method validation package and comparability protocols. In an EU submission, this should contain a process validation scheme for the drug product, specific forms relating to the prevention of TSEs from animal-derived materials or other infections from materials of human origin, relevant certificates of suitability and Medical Device information if such is used as part of the drug delivery system.

6.3.3
Drug Master Files

It should be noted that all technical data submitted in support of marketing authorisations remain confidential between the applicant and the regulatory authorities. In situations where the manufacturer of a drug substance is a different commercial entity to the applicant seeking approval of a drug product incorporating the substance, the substance manufacturer may submit quality data on the substance directly to the authorities as a Drug Master File. This enables the authorities to review information on the quality of the drug substance as part of the overall review of the product marketing application, without requiring the substance manufacturer to disclose his/her manufacturing procedures to the marketing applicant. However, the substance manufacturer must provide the substance user with all the relevant procedures necessary to assess the quality of the substance before it is incorporated into the drug product.

6.3.4
Module 4: Non-Clinical Study Reports

This module should contain data from all of the non-clinical studies conducted to investigate the pharmacological and toxicological properties of the drug substance/product. The standard headings used to present this information are shown in Figure 6.4. Study reports should be presented in the following order according to species and route of administration:

Species	Route of administration
Mouse	Oral
Rat	Intravenous
Hamster	Intramuscular
Other rodent	Intraperitoneal
Rabbit	Subcutaneous
Dog	Inhalation
Non-human primate	Topical
Other non-rodent mammal	Other
Non-mammals	

6.3.5
Module 5: Clinical Study Reports

This module should contain reports of all the clinical studies and other related data that were conducted to demonstrate the safety and efficacy of the drug in human subjects. The standard headings used to present this information are shown in Figure 6.5. An example of headings for a tabulated listing of clinical studies is shown in Table 6.4.

The following points are worthy of note in terms of the placement of data. In the case of studies with multiple objectives, reports should be placed in the section corresponding to their primary purpose. Reports of laboratory studies conducted with human materials to investigate pharmacokinetic effects should be placed in Section 5.3.2 of the clinical module, as opposed to the non-clinical module. A US submission requires that the individual case report forms of all trial subjects that died or were dropped from a study due to adverse events are included in Section 5.3.7.

4.1 **Table of Contents of Module 4**

4.2 **Study Reports**

4.2.1 **Pharmacology**
4.2.1.1 Primary Pharmacodynamics
4.2.1.2 Secondary Pharmacodynamics
4.2.1.3 Safety Pharmacology
4.2.1.4 Pharmacodynamic Drug Interactions

4.2.2 **Pharmacokinetics**
4.2.2.1 Analytical Methods and Validation Reports
4.2.2.2 Absorption
4.2.2.3 Distribution
4.2.2.4 Metabolism
4.2.2.5 Excretion
4.2.2.6 Pharmacokinetic Drug Interactions (non-clinical)
4.2.2.7 Other Pharmacokinetic Studies

4.2.3 **Toxicology**
4.2.3.1 Single-Dose Toxicity
4.2.3.2 Repeat-Dose Toxicity
4.2.3.3 Genotoxicity
    4.2.3.3.1 *In vitro*
    4.2.3.3.2 *In vivo*
4.2.3.4 Carcinogenicity
    4.2.3.4.1 Long-term studies
    4.2.3.4.2 Short- or medium-term studies
    4.2.3.4.3 Other studies
4.2.3.5 Reproductive and Developmental Toxicity
    4.2.3.5.1 Fertility and early embryonic development
    4.2.3.5.2 Embryo-fetal development
    4.2.3.5.3 Prenatal and postnatal development, including maternal function
    4.2.3.5.4 Studies in which the offspring (juvenile animals) are dosed and/or further evaluated.
4.2.3.6 Local Tolerance
4.2.3.7 Other Toxicity Studies (if available)
    4.2.3.7.1 Antigenicity
    4.2.3.7.2 Immunotoxicity
    4.2.3.7.3 Mechanistic studies (if not included elsewhere)
    4.2.3.7.4 Dependence
    4.2.3.7.5 Metabolites
    4.2.3.7.6 Impurities
    4.2.3.7.7 Other

4.3 **Literature References**

Figure 6.4 The headings structure of Module 4.

Table of Contents of Module 5

5.2   Tabular Listing of All Clinical Studies

5.3   Clinical Study Reports

5.3.1 Reports of Biopharmaceutic Studies
     5.3.1.1   Bioavailability (BA) Study Reports
     5.3.1.2   Comparative BA and Bioequivalence (BE) Study Reports
     5.3.1.3   *In vitro-In vivo* Correlation Study Reports
     5.3.1.4   Reports of Bioanalytical and Analytical Methods for Human Studies

5.3.2 Reports of Studies Pertinent to Pharmacokinetics using Human Biomaterials
     5.3.2.1   Plasma Protein Binding Study Reports
     5.3.2.2   Reports of Hepatic Metabolism and Drug Interaction Studies
     5.3.2.3   Reports of Studies Using Other Human Biomaterials

5.3.3 Reports of Human Pharmacokinetic (PK) Studies
     5.3.3.1   Healthy Subject PK and Initial Tolerability Study Reports
     5.3.3.2   Patient PK and Initial Tolerability Study Reports
     5.3.3.3   Intrinsic Factor PK Study Reports (sex, age, race, weight, etc.)
     5.3.3.4   Extrinsic Factor PK Study Reports (diet, smoking, alcohol use, etc.)
     5.3.3.5   Population PK Study Reports

5.3.4 Reports of Human Pharmacodynamic (PD) Studies
     5.3.4.1   Healthy Subject PD and PK/PD Study Reports
     5.3.4.2   Patient PD and PK/PD Study Reports

5.3.5 Reports of Efficacy and Safety Studies
     5.3.5.1   Study Reports of Controlled Clinical Studies Pertinent to the Claimed Indication
     5.3.5.2   Study Reports of Uncontrolled Clinical Studies
     5.3.5.3   Reports of Analyses of Data from More Than One Study
     5.3.5.4   Other Clinical Study Reports

5.3.6 Reports of Post-Marketing Experience

5.3.7 Case Report Forms and Individual Patient Listings

5.4   Literature References

Figure 6.5 The headings structure of Module 5.

Copies of referenced documents, including important published articles, official meeting minutes, or other regulatory guidance or advice should be provided in the reference section.

6.3.6
Module 2: Summaries

This module should be prefaced by a brief introduction outlining the pharmacological class, mode of action and proposed clinical use of the drug. The overview reports form the most important part of this module as they introduce the reviewer to the case that the applicant wishes to put forward in favour of authorisation of the drug for the

Table 6.4 Example of a tabulated listing of clinical studies.

Type of Study	Study Identifier	Location of Study Report	Objective(s) of the Study	Study Design and Type of Control	Test Product(s); Dosage Regimen; Route of Administration	Number of Subjects	Healthy Subjects or Diagnosis of Patients	Duration of Treatment	Study Status; Type of Report
BA	001	Vol. 3, Sec. 1.1, p. 183	Absolute BA IV vs. Tablet	Cross-over	Tablet, 50 mg single dose, oral, 10 mg i.v.	20	Healthy subjects	Single dose	Complete; Abbreviated
BE	002	Vol. 4, Sec. 1.2, p. 254	Compare clinical study and to-be-marketed formulation	Cross-over	Two tablet formulations, 50 mg, oral	32	Healthy subjects	Single dose	Complete; Abbreviated
PK	1010	Vol. 6, Sec. 3.3, p. 29	Define PK	Cross-over	Tablet, 50 mg single dose, oral	50	Renal insufficiency	Single dose	Complete; Full
PD	020	Vol. 6, Sec. 4.2, p. 147	Bridging study between regions	Randomised placebo-controlled	Tablet, 50 mg, multiple dose, oral, every 8 h	24 (12 drug, 12 placebo)	Patients with primary hypertension	2 weeks	Ongoing; Interim
Efficacy	035	Vol. 10, Sec. 5.1, p. 1286	Long term; Efficacy and Safety; Population PK analysis	Randomised active-controlled	Tablet, 50 mg, oral, every 8 h	300 (152 test drug, 148 active control)	Patients with primary hypertension	48 weeks	Complete; Full

BA = Bioavailability, BE = Bioequivalence, PK = Pharmacokinetic, PD = Pharmacodynamic, IV/i.v. = Intravenous.

proposed indications for use. They should be written by experts, and provide reasoned and integrated evaluations and assessments of the data. As such, it is expected that the authors will draw on data from across different studies to substantiate particular points in terms of efficacy, safety, or quality. Similarly, they should highlight any limitations or areas of concern that may exist. Quality aspects should be covered in a single Quality Overall Summary document, that summarises and justifies the quality measures specified. This should follow the structure of Module 3. The reviews of the non-clinical and clinical aspects are each split into two sections. The non-clinical overview and clinical overview sections should present critical discussions and assessments of results with conclusions, whereas the non-clinical and clinical summaries provide abbreviated factual data presented as written summaries and supported by summary tables. The structure of these sections is shown in Figure 6.6.

Non-Clinical Overview
Overview of the non-clinical testing strategy
Pharmacology
Pharmacokinetics
Toxicology
Integrated overview and conclusions
List of literature references

Non-Clinical Summaries
Introduction
Written Summary of Pharmacology
Tabulated Summary of Pharmacology
Written Summary of Pharmacokinetics
Tabulated Summary of Pharmacokinetics
Written Summary of Toxicology
Tabulated Summary of Toxicology

Clinical Overview
Product Development Rationale
Overview of Biopharmaceutics
Overview of Clinical Pharmacology
Overview of Efficacy
Overview of Safety
Benefits and Risks Conclusions
Literature References

Clinical Summaries
Summary of Biopharmaceutic Studies and Associated Analytical Methods
Summary of Clinical Pharmacology Studies
Summary of Clinical Efficacy
Summary of Clinical Safety
Literature References
Synopses of Individual Studies

Figure 6.6 Structure of the Non-clinical and Clinical Overview and Summary sections.

6.3.7
Module I: Region-Specific

The content of information submitted in this module will vary depending on the requirements of the different regional authorities. We shall examine some of the specific features that apply to EU and US submissions.

6.3.8
Module 1: EU

The headings for an EU submission are shown in Figure 6.7. Specific applications forms, which are available on the Eduralex website, may be completed either electronically or in hard copy. A summary of the information captured by the form is shown in Figure 6.8.

EU regulations require the submission of a Summary of Product Characteristics (SPC) as part of the product information section. Specific templates and guidance are

1.0 Cover Letter
1.1 Comprehensive Table of Contents
1.2 Application Form
1.3 Product Information
    1.3.1 SPC, Labelling and Package Leaflet
    1.3.2 Mock-up
    1.3.3 Specimen
    1.3.4 Consultation with Target Patient Groups
    1.3.5 Product Information already approved in the Member States
    1.3.6 Braille
1.4 Information about the Experts
    1.4.1 Quality
    1.4.2 Non-Clinical
    1.4.3 Clinical
1.5 Specific Requirements for Different Types of Applications
    1.5.1 Information for Bibliographical Applications
    1.5.2 Information for Generic, 'Hybrid' or Bio-similar Applications
    1.5.3 (Extended) Data/Market Exclusivity
    1.5.4 Exceptional Circumstances
    1.5.5 Conditional Marketing Authorisation
1.6 Environmental Risk Assessment
    1.6.1 Non-GMO
    1.6.2 GMO
1.7 Information relating to Orphan Market Exclusivity
    1.7.1 Similarity
    1.7.2 Market Exclusivity
1.8 Information relating to Pharmacovigilance
    1.8.1 Pharmacovigilance System
    1.8.2 Risk-management System
1.9 Information relating to Clinical Trials

Figure 6.7 Module 1 headings: EU submission.

1. **Type of Application**
1.1   Application Type by Authorisation Procedure
  1.1.1 Centralised Procedure
  1.1.2 Mutual Recognition Procedure
  1.1.3 Decentralised Procedure
  1.1.4 National procedure
1.2   Orphan Product Information
  1.2.1 Orphan designation status
  1.2.2 Market exclusivity
1.3   Marketing Extension Application (change in active substance, bioavailability, pharmacokinetics, strength, pharmaceutical form or route of administration, if applicable)
1.4   Application Type by supporting data
  1.4.1. Standard application (Quality, Safety, Efficacy)
  1.4.2. Generic Application (based on reference product Safety and Efficacy data)
  1.4.3. Hybrid Application (variation on a reference product requiring additional Safety and Efficacy data)
  1.4.4. Similar Biological Product Application
  1.4.5. Well Established Use Application (bibliographic)
  1.4.6. Fixed Combination Product Application
  1.4.7. Informed Consent Application
  1.4.8. Traditional Use Registration of a Herbal Product
1.5   Special Condition Requests (conditional approval, accelerated review, additional year data exclusivity)

2. **Marketing Authorisation Application Particulars**
  2.1 Names and ATC code of drug
  2.2 Strength, form route of administration, container and pack sizes
  2.3 Legal status (non-prescription or prescription only, renewable/non-renewable/special/restricted)
  2.4 Marketing Authorisation Holder (identity and contacts)
  2.5 Manufacturers (contacts, inspection status)
  2.6 Qualitative and Quantitative Composition

3. **Scientific Advice**
    (Received from CHMP or CAs)

4. **Paediatric Development Programme**

5. **Other Marketing Authorisation Applications**
    (Applications or authorisations in other Member States or outside the EU)

6. **Annexed Documents**
    (Proof of payment, rights of reference letters, flow charts of production sites, manufacturing authorisations, GMP certificates, etc.)

Figure 6.8 Outline of information captured by an EU Marketing Authorisation Application form.

available to prepare the document, which must adhere to the headings shown in Figure 6.9. The SPC is a pivotal document in terms of how the medicinal product can be presented and promoted. The authorities must approve the SPC as part of the granting of a marketing authorization, and thus it will usually require amendment

1. Name, strength and pharmaceutical form

2. Qualitative and Quantitative composition

3. Pharmaceutical form

4. Clinical Particulars: (therapeutic indications, administration, contraindications, special warnings and precautions, interaction with other drugs, use in pregnancy or lactation, effects on driving, undesirable effects, overdose)

5. Pharmacological Properties: (pharmacodynamics, pharmacokinetics, pre-clinical safety data)

6. Pharmaceutical particulars: (excipients, major incompatibilities, shelf life, storage precautions, nature and contents of container, disposal precautions)

7. Marketing authorisation holder.

8. Marketing authorisation number(s).

9. Date of the first authorisation or renewal of the authorisation.

10. Date of revision of the text.

Figure 6.9 Summary of Product Characteristics (SPC) headings.

over the course of the review process so that a final agreed version can be authorised as part of the licensing decision.

The package leaflet and label or other information provided to health professionals on the safe and effective use of the product must be drawn up in accordance with the SPC. The information that must be provided in the package leaflet and outer package label are shown in Figures 6.10 and 6.11, respectively. The languages used should be appropriate to the locations where authorizations are sought, and target groups should be consulted to ensure that text is legible, clear and easy to use. Specific approval of label and leaflet text is an integral part of the authorisation process. Additional non-harmonised information that individual member states may require, such as price, reimbursement conditions, legal status and identification/authenticity should be placed in a boxed area known as the "blue" box.

Section 1.4 must contain the signatures and information on the credentials of the experts who wrote the Module 2 summaries, as this is a specific legal requirement of

Identification of the medicinal product: (name, form, strength, pharmaco-therapeutic group)

Therapeutic indications

Information before use: (contraindications, precautions, interactions with drugs/alcohol, warnings)

Usage instructions: (dosage, method of taking, frequency, duration, overdose, missed dose)

Adverse reactions

Information on: (usage within expiry date, storage, signs of deterioration, qualitative composition, pharmaceutical form and content, name and address of marketing authorisation holder and manufacturer)

Additional names under which the product is authorised in other member states, if applicable.

Date of last revision of leaflet

Figure 6.10 Information to be provided in the package leaflet.

Name form and strength of medicinal product

Active substance per dosage

Quantity by weight, volume or dosage

Excipients (if required)

Method of administration

Warning to keep out of reach of children

Special warnings, if necessary

Expiry date

Special storage conditions

Special precautions about disposal, if applicable

Name and address of the marketing authorisation holder

Marketing authorisation number

Batch number

Instructions for use if non-prescription

Figure 6.11 Information to be provided in the outer package label.

the EU system. The applicant must submit an environmental risk assessment, evaluating any potential risks to the environment arising from use, storage and disposal of medicinal products. In the case of products containing genetically modified organisms (GMOs), the information must be presented in accordance with the provisions of Directive 2001/18/EC on the deliberate release into the environment of GMOs.

6.3.9
Module 1: US

The contents of a US submission are shown in Figure 6.12. A single two-page application form covers all drug submissions for human use, whether they are a New Drug Application (NDA), a Biologics License Application (BLA), or an Abbreviated New Drug Application (ANDA). The form should be used with each submission communication, and serves as a submission checklist and a commitment to comply with applicable regulations and other requirements. A sample is shown in Figure 6.13. Certification that a field copy has been sent to the local FDA office (for use during inspections of the manufacturing facilities) is only necessary if making a paper submission. Financial disclosure information on investigators must be submitted on specific forms. Environmental assessments can be avoided if it is estimated that the active ingredient will not reach 1 part per billion in the aquatic environment after the fifth year of marketing. The prescribing information consists of copies of the labelling and package insert, plus a Medication Guide applicable to prescription drugs that may be taken without direct medical supervision. The specific labelling requirements are set out in 21 CFR Parts, 201, 606, 610 and 660. The annotated

1. FDA form 356h
2. Comprehensive table of contents
3. Administrative documents
   a. Administrative documents
      Patent information on any patent that claims the drug, if applicable
      Patent certifications (not for BLA)
      Debarment certification
      Field copy certification (not for BLA)
      User fee cover sheet
      Financial disclosure information
      Letters of authorization for reference to other applications or drug master files
      Waiver requests
      Environmental assessment or request for categorical exclusion
      Statements of claimed exclusivity and associated certifications
   b Prescribing information
   c. Annotated labeling text
   d. Labeling comparison (for ANDA)

Figure 6.12 Module 1 contents: US submission.

version of the labelling should show annotations cross-referencing to the data that substantiate each of the statements contained in the labelling.

6.4
Submission and Review Process in the EU

Depending on the type of medicinal product, different procedures exist for obtaining a marketing authorisation. "Community Authorisations" are granted via a "centralised" procedure in which applications are submitted to the European Medicines Agency, whereas "National Authorisations" may be obtained by applying to the Competent Authorities of member states using either "decentralised" or "mutual recognition" procedures.

6.4.1
Community Authorisation

The procedures for obtaining a Community marketing authorisation are defined in EC Regulation No. 726/2004. The types of human-use medicinal products for which the procedure may be used are shown in Figure 6.14. The applicant should notify the EMEA of their intention to submit an application at least 7 months

DEPARTMENT OF HEALTH AND HUMAN SERVICES FOOD AND DRUG ADMINISTRATION	Form Approved: OMB No. 0910-0430 Expiration Date: April 30, 2009 See OMB Statement on page 2.
APPLICATION TO MARKET A NEW DRUG, BIOLOGIC, OR AN ANTIBIOTIC DRUG FOR HUMAN USE *(Title 21, Code of Federal Regulations, Parts 314 & 601)*	**FOR FDA USE ONLY** APPLICATION NUMBER

APPLICANT INFORMATION

NAME OF APPLICANT	DATE OF SUBMISSION
TELEPHONE NO. *(Include Area Code)*	FACSIMILE (FAX) Number *(Include Area Code)*
APPLICANT ADDRESS *(Number, Street, City, State, Country, ZIP Code or Mail Code, and U.S. License number if previously issued):*	AUTHORIZED U.S. AGENT NAME & ADDRESS *(Number, Street, City, State, ZIP Code, telephone & FAX number)* IF APPLICABLE

PRODUCT DESCRIPTION

NEW DRUG OR ANTIBIOTIC APPLICATION NUMBER, OR BIOLOGICS LICENSE APPLICATION NUMBER *(If previously issued)*

ESTABLISHED NAME *(e.g., Proper name, USP/USAN name)*	PROPRIETARY NAME *(trade name)* IF ANY
CHEMICAL/BIOCHEMICAL/BLOOD PRODUCT NAME *(If any)*	CODE NAME *(If any)*

DOSAGE FORM:	STRENGTHS:	ROUTE OF ADMINISTRATION:

(PROPOSED) INDICATION(S) FOR USE:

APPLICATION INFORMATION

APPLICATION TYPE *(check one)* ☐ NEW DRUG APPLICATION (NDA, 21 CFR 314.50) ☐ ABBREVIATED NEW DRUG APPLICATION (ANDA, 21 CFR 314.94)
☐ BIOLOGICS LICENSE APPLICATION (BLA, 21 CFR Part 601)

IF AN NDA, IDENTIFY THE APPROPRIATE TYPE ☐ 505 (b)(1) ☐ 505 (b)(2)

IF AN ANDA, OR 505(b)(2), IDENTIFY THE REFERENCE LISTED DRUG PRODUCT THAT IS THE BASIS FOR THE SUBMISSION
Name of Drug   Holder of Approved Application

TYPE OF SUBMISSION *(check one)* ☐ ORIGINAL APPLICATION ☐ AMENDMENT TO A PENDING APPLICATION ☐ RESUBMISSION
☐ PRESUBMISSION ☐ ANNUAL REPORT ☐ ESTABLISHMENT DESCRIPTION SUPPLEMENT ☐ EFFICACY SUPPLEMENT
☐ LABELING SUPPLEMENT ☐ CHEMISTRY MANUFACTURING AND CONTROLS SUPPLEMENT ☐ OTHER

IF A SUBMISSION OF PARTIAL APPLICATION, PROVIDE LETTER DATE OF AGREEMENT TO PARTIAL SUBMISSION: _____

IF A SUPPLEMENT, IDENTIFY THE APPROPRIATE CATEGORY ☐ CBE ☐ CBE-30 ☐ Prior Approval (PA)

REASON FOR SUBMISSION

PROPOSED MARKETING STATUS *(check one)* ☐ PRESCRIPTION PRODUCT (Rx) ☐ OVER THE COUNTER PRODUCT (OTC)

NUMBER OF VOLUMES SUBMITTED _____ THIS APPLICATION IS ☐ PAPER ☐ PAPER AND ELECTRONIC ☐ ELECTRONIC

ESTABLISHMENT INFORMATION (Full establishment information should be provided in the body of the application.)
Provide locations of all manufacturing, packaging and control sites for drug substance and drug product (continuation sheets may be used if necessary). Include name, address, contact, telephone number, registration number (CFN), DMF number, and manufacturing steps and/or type of testing (e.g. Final dosage form, Stability testing) conducted at the site. Please indicate whether the site is ready for inspection or, if not, when it will be ready.

Cross References (list related License Applications, INDs, NDAs, PMAs, 510(k)s, IDEs, BMFs, and DMFs referenced in the current application)

FORM FDA 356h (4/06)   PSC Graphics: (301) 443-1090 EF
                                                    PAGE 1

Figure 6.13 US application form FDA 356 h.

This application contains the following items: *(Check all that apply)*

☐	1.	Index
☐	2.	Labeling *(check one)* ☐ Draft Labeling ☐ Final Printed Labeling
☐	3.	Summary (21 CFR 314.50 (c))
☐	4.	Chemistry section
		A. Chemistry, manufacturing, and controls information (e.g., 21 CFR 314.50(d)(1); 21 CFR 601.2)
		B. Samples (21 CFR 314.50 (e)(1); 21 CFR 601.2 (a)) (Submit only upon FDA's request)
		C. Methods validation package (e.g., 21 CFR 314.50(e)(2)(i); 21 CFR 601.2)
☐	5.	Nonclinical pharmacology and toxicology section (e.g., 21 CFR 314.50(d)(2); 21 CFR 601.2)
☐	6.	Human pharmacokinetics and bioavailability section (e.g., 21 CFR 314.50(d)(3); 21 CFR 601.2)
☐	7.	Clinical Microbiology (e.g., 21 CFR 314.50(d)(4))
☐	8.	Clinical data section (e.g., 21 CFR 314.50(d)(5); 21 CFR 601.2)
☐	9.	Safety update report (e.g., 21 CFR 314.50(d)(5)(vi)(b); 21 CFR 601.2)
☐	10.	Statistical section (e.g., 21 CFR 314.50(d)(6); 21 CFR 601.2)
☐	11.	Case report tabulations (e.g., 21 CFR 314.50(f)(1); 21 CFR 601.2)
☐	12.	Case report forms (e.g., 21 CFR 314.50 (f)(2); 21 CFR 601.2)
☐	13.	Patent information on any patent which claims the drug (21 U.S.C. 355(b) or (c))
☐	14.	A patent certification with respect to any patent which claims the drug (21 U.S.C. 355 (b)(2) or (j)(2)(A))
☐	15.	Establishment description (21 CFR Part 600, if applicable)
☐	16.	Debarment certification (FD&C Act 306 (k)(1))
☐	17.	Field copy certification (21 CFR 314.50 (l)(3))
☐	18.	User Fee Cover Sheet (Form FDA 3397)
☐	19.	Financial Information (21 CFR Part 54)
☐	20.	OTHER *(Specify)*

CERTIFICATION

I agree to update this application with new safety information about the product that may reasonably affect the statement of contraindications, warnings, precautions, or adverse reactions in the draft labeling. I agree to submit safety update reports as provided for by regulation or as requested by FDA. If this application is approved, I agree to comply with all applicable laws and regulations that apply to approved applications, including, but not limited to the following:

1. Good manufacturing practice regulations in 21 CFR Parts 210, 211 or applicable regulations, Parts 606, and/or 820.
2. Biological establishment standards in 21 CFR Part 600.
3. Labeling regulations in 21 CFR Parts 201, 606, 610, 660, and/or 809.
4. In the case of a prescription drug or biological product, prescription drug advertising regulations in 21 CFR Part 202.
5. Regulations on making changes in application in FD&C Act section 506A, 21 CFR 314.71, 314.72, 314.97, 314.99, and 601.12.
6. Regulations on Reports in 21 CFR 314.80, 314.81, 600.80, and 600.81.
7. Local, state and Federal environmental impact laws.

If this application applies to a drug product that FDA has proposed for scheduling under the Controlled Substances Act, I agree not to market the product until the Drug Enforcement Administration makes a final scheduling decision.

The data and information in this submission have been reviewed and, to the best of my knowledge are certified to be true and accurate.

Warning: A willfully false statement is a criminal offense, U.S. Code, title 18, section 1001.

SIGNATURE OF RESPONSIBLE OFFICIAL OR AGENT	TYPED NAME AND TITLE	DATE

ADDRESS *(Street, City, State, and ZIP Code)*	Telephone Number

Public reporting burden for this collection of information is estimated to average 24 hours per response, including the time for reviewing instructions, searching existing data sources, gathering and maintaining the data needed, and completing and reviewing the collection of information. Send comments regarding this burden estimate or any other aspect of this collection of information, including suggestions for reducing this burden to:

Department of Health and Human Services
Food and Drug Administration
Center for Drug Evaluation and Research (HFD-143)
Central Document Room
5901-B Ammendale Road
Beltsville, MD 207052-1266

Department of Health and Human Services
Food and Drug Administration
Center for Biologics Evaluation and Research (HFM-99)
1401 Rockville Pike
Rockville, MD 20852-1448

An agency may not conduct or sponsor, and a person is not required to respond to, a collection of information unless it displays a currently valid OMB control number.

FORM FDA 356h (4/06)

PAGE 2

Figure 6.13 *(Continued)*

INSTRUCTIONS FOR FILLING OUT FORM FDA 356h

APPLICANT INFORMATION This section should include the name, street address, telephone, and facsimile numbers of the legal person or entity submitting the application in the appropriate areas. Note that, in the case of biological products, this is the name of the legal entity or person to whom the license will be issued. The name, street address and telephone number of the legal person or entity authorized to represent a non-U.S. applicant should be entered in the indicated area. Only one person should sign the form.

PRODUCT DESCRIPTION This section should include all of the information necessary to identify the product that is the subject of this submission. For new applications, the proposed indication should be given. For supplements to an approved application, please give the approved indications for use.

APPLICATION INFORMATION If this submission is an ANDA or 505(b)(2), this section should include the name of the approved drug that is the basis of the application and identify the holder of the approved application in the indicated areas.

TYPE OF SUBMISSION should be indicated by checking the appropriate box:

Original Application = a complete new application that has never before been submitted;

Amendment to a Pending Application = all submissions to pending original applications, or pending supplements to approved applications, including responses to Information Request Letters;

Resubmission = a complete response to an action letter, or submission of an application that has been the subject of a withdrawal or a refusal to file action;

Presubmission = information submitted prior to the submission of a complete new application;

Annual Report = periodic reports for licensed biological products (for NDAs Form FDA-2252 should be used as required in 21 CFR 314.81 (b)(2));

Establishment Description Supplement = supplements to the information contained in the Establishment Description section (#15) for biological products;

Efficacy Supplement = submissions for such changes as a new indication or dosage regimen for an approved product, a comparative efficacy claim naming another product, or a significant alteration in the patient population; e.g., prescription to Over-The-Counter switch;

Labeling Supplement = all label change supplements required under 21 CFR 314.70 and 21 CFR 601.12 that do not qualify as efficacy supplements;

Chemistry, Manufacturing, and Controls Supplement = manufacturing change supplement submissions as provided in 21 CFR 314.70, 21 CFR 314.71, 21 CFR 314.72 and 21 CFR 601.12;

Other = any submission that does not fit in one of the other categories (e.g., Phase IV response). If this box is checked the type of submission can be explained in the **REASON FOR SUBMISSION** block.

Submission of Partial Application Letter date of agreement to partial submission should be provided. Also, provide copy of scheduled plan.

CBE "Supplement-Changes Being Effected" supplement submission for certain moderate changes for which distribution can occur when FDA receives the supplement as provided in 21 CFR 314.70 and 21 CFR 601.12.

Figure 6.13 *(Continued)*

CBE-30 "Supplement-Changes Being Effected in 30 Days" supplement submission for certain moderate changes for which FDA receives at least 30 days before the distribution of the product made using the change as provided in 21 CFR 314.70 and 21 CFR 601.12.

Prior Approval (PA) "Prior Approval Supplements" supplement submission for a major change for which distribution of the product made using the change cannot occur prior to FDA approval as provided in 21 CFR 314.70 and 21 CFR 601.12.

REASON FOR SUBMISSION This section should contain a brief explanation of the submission, e.g., "manufacturing change from roller bottle to cell factory" or "response to Information Request Letter of 1/9/97" or "Pediatric exclusivity determination request" or "to satisfy a subpart H postmarketing commitment".

NUMBER OF VOLUMES SUBMITTED Please enter the number of volumes, including and identifying electronic media, contained in the archival copy of this submission.

This application is
☐ Paper ☐ Paper and Electronic ☐ Electronic
Please check the appropriate box to indicate whether this submission contains only paper, both paper and electronic media, or only electronic media.

ESTABLISHMENT INFORMATION This section should include information on the locations of all manufacturing, packaging and control sites for both drug substance and drug product. If continuation sheets are used, please indicate where in the submission they may be found. For each site please include the name, address, telephone number, registration number (Central File Number), Drug Master File (DMF) number, and the name of a contact at the site. The manufacturing steps and/or type of testing (e.g. final dosage form, stability testing) conducted at the site should also be included. Please indicate whether the site is ready for inspection or, if not, when it will be ready. Please note that, when applicable, the complete establishment description is requested under item 15.

CROSS REFERENCES This section should contain a list of all License Applications, Investigational New Drug Applications (INDs), NDAs, Premarket Approval Applications (PMAs), Premarket Notifications (510(k)s), Investigational Device Exemptions (IDEs), Biological Master Files (BMFs) and DMFs that are referenced in the current application.

Items 1 through 20 on the reverse side of the form constitute a check list that should be used to indicate the types of information contained within a particular submission. Please check all that apply. The numbering of the items on the checklist is not intended to specify a particular order for the inclusion of those sections into the submission. The applicant may include sections in any order, but the location of those sections within the submission should be clearly indicated in the Index. It is therefore recommended that, particularly for large submissions, the Index immediately follows the Form FDA 356h and, if applicable, the User Fee Cover Sheet (Form FDA 3397).

The CFR references are provided for most items in order to indicate what type of information should be submitted in each section. For further information, the applicant may consult the guidance documents that are available from the Agency.

Signature The form must be signed and dated. Ordinarily only one person should sign the form, i.e., the applicant, or the applicant's attorney, agent, or other authorized official. However, if the person signing the application does not reside or have a place of business within the United States, the application should be countersigned by an attorney, agent, or other authorized official who resides or maintains a place of business within the United States.

FORM FDA 356h (4/06)

Figure 6.13 (*Continued*)

Mandatory

- Biotech products based on recombinant DNA, gene expression or hybridoma/monoclonal antibody technologies.
- New active substances for treatment of AIDS, cancer, diabetes, neurodegenerative disorders, (and from 2008) autoimmune or other immunological disorders and viral diseases.
- Orphan medicinal products.

Optional

- Other new active substances.
- Medicinal products that contribute significant therapeutic, scientific or technical innovation, or are in the interests of patient health.
- A generic copy of a centrally authorised product.

Figure 6.14 Human medicinal products that may be authorised via the centralised procedure.

in advance of the planned submission date. This pre-submission should contain a draft of the SPC, the grounds on which a Community Authorisation will be sought, a proposed invented name that will identify the drug throughout the Community, and other information relevant to the proposed application strategy. Applicants are also encouraged to have a pre-submission meeting with the EMEA to discuss any aspects of the application in greater detail. The pre-submission notification allows the Committee for Medicinal Products for Human Use (CHMP) to verify that the product is suitable for authorisation via the centralised procedure, and to appoint a Rapporteur and Co-rapporteur from among its members. The rapporteurs are responsible for conducting the scientific evaluation of the dossier, and will assemble evaluation teams drawn from the panel of scientific experts. In practice, they will rely heavily on the resources available through their respective Competent Authorities. An EMEA product team is organised in parallel to deal with all procedural aspects of the application. Currently, the fees for a complete application of a single strength product as of December 2007 are €232 000.

6.4.2
Scientific Evaluation Process

The applicant must submit complete copies of dossier to the EMEA, and directly to each of the rapporteurs. The EMEA are allowed 14 days to validate the application to ensure that it is complete and in accordance with the regulations, after which the rapporteurs can commence the scientific evaluation. The review process involves considerable feedback between the rapporteurs, the CHMP and the applicant, central to which can be achieving agreement on the final text of the SPC, labels and leaflet. The rapporteurs are required to deliver preliminary assessment reports to the CHMP within 80 days of the start of the review process. The applicant is also copied on the report. Over the next 40 days issues raised in the reports, together with the comments and feedback from the CHMP, are formalised into a list of questions that are officially sent to the applicant. The CHMP may additionally request that GMP/GCP/GLP inspections are conducted by the Competent Authorities of the relevant Member States, or that drug samples are tested. The clock is then stopped until the applicant provides responses to these questions. The rapporteurs have another 30 days to prepare a joint report on these responses. A further 30 days are allowed for the CHMP to issue a final list of any outstanding issues and/or a request for an oral presentation by the applicant. If an oral presentation is necessary, the clock stops until this is complete. Otherwise, the applicant must submit the final draft of the SPC, label and leaflet text in English within the next 30 days. At this stage, the CHMP will adopt a formal scientific opinion on the application. The total time allowed to compete the scientific review is 210 days, excluding the time taken for applicant responses.

Assuming that the outcome is favourable, the agency must forward its opinion to the applicant, the EU Commission and the Member States within 15 days, together with:

- A report describing the evaluation of the medicinal product and the reasoning behind the conclusions drawn.
- Copies of the draft SPC (this becomes Annex 1 to the decision).
- Manufacturing information and any proposed restrictions or conditions (Annex 2)
- The draft text of the label and package leaflet (Annex 3).
- The Assessment Report.

If the opinion is unfavourable, the applicant has 15 days to notify the agency of its intention to request a re-examination of the decision, and a further 45 days to submit the grounds on which a review is requested. New rapporteurs are appointed to re-examine the dossier, and the CHMP is allowed 60 days to deliver its opinion in light of the re-examination.

6.4.3
Decision-Making Process

The Commission are responsible for turning the agency opinion into a binding decision for all Member States, and ensuring that the marketing authorisation is in compliance with the regulations. They must prepare a draft decision in consultation with other relevant Directorates General within 15 days of receipt of the opinion from the EMEA. In the meantime, the EMEA is responsible for coordinating with the Member States, Norway and Iceland, a linguist review of translations of the SPC, label and leaflet into EU, Norwegian and Icelandic languages, prepared by the applicant Note: Norway, Iceland and Liechtenstein, through the European Economic Area (EEA) agreement, have adopted the EU system for drug regulation, even though they are not EU member states. The states have 15 days to return linguist comments, so that the Commission may annex final text versions to the draft decision before it is forwarded to the members of the Standing Committee for Medicinal Products for

Human Use for consideration. The Committee should return its comments and observations in writing to the Commission within 22 days, unless it raises serious objections on legal or public health grounds that necessitate discussion in a plenary meeting of the Committee. Once the Committee consultation stage has been completed, the Commission has 15 days to adopt the final decision. The decision is signed by the Director General of the Directorate General for Enterprise & Industry; the applicant and all Member States are notified, and the decision is published in the next issue of the Official Journal of the EU, C series. A European Public Assessment Report (EPAR) is prepared by the CHMP and published on the EMEA website. This is intended to inform the general public in simple, easy-to-understand language of the reasoning behind the granting or refusal of a marketing authorisation. Excluding the time taken for applicant responses or appeals, a Community marketing authorisation should be granted within 277 days of the start of the review process. A flow diagram of the authorisation process is shown in Figure 6.15.

6.4.4
National Authorisations

Outside of the products that come within the scope of Community Authorisations, all other products can only be licensed via application to the Competent Authorities of individual Member States. However, through the use of either decentralised or mutual recognition procedures it is possible to obtain authorisations on the basis of a dossier assessment conducted by a single Member State.

6.4.5
Decentralised Procedure

The decentralised procedure can be used in cases where the product has never been authorised in any of the Member States, and the applicant wishes to obtain a license in a number of states simultaneously. The applicant must submit applications with the complete dossier to the Competent Authorities of each of the Member States where authorisation is desired. A single Member State should be chosen as the "reference" state to undertake the scientific assessment of the complete dossier, while the other states are designated as "concerned" states. The review process has many parallels with the centralised procedure, in that similar time lines exist, the reference State plays the role of the rapporteur, and the concerned States replace the CHMP. Once all States have validated that the dossiers are complete, the reference State is allowed 70 days to review the dossier and prepare a preliminary assessment report, which is circulated to the concerned Member States and the applicant. Comments from the concerned Member States and applicant responses are collected so that by day 120 the reference State may issue a draft assessment report together with draft SPC, label and leaflet texts. The clock may be stopped until requested responses from the applicant are received. The application then enters the second step in the assessment process, during which all the concerned Member States consider the

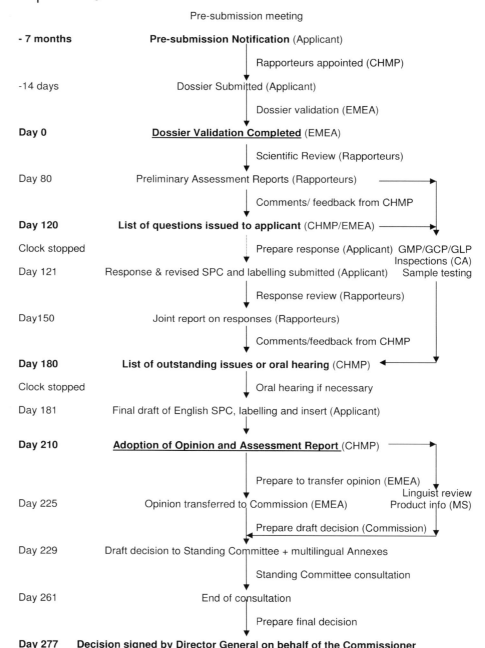

Pre-submission meeting

- 7 months   **Pre-submission Notification** (Applicant)

Rapporteurs appointed (CHMP)

-14 days   Dossier Submitted (Applicant)

Dossier validation (EMEA)

Day 0   **Dossier Validation Completed** (EMEA)

Scientific Review (Rapporteurs)

Day 80   Preliminary Assessment Reports (Rapporteurs)

Comments/ feedback from CHMP

Day 120   **List of questions issued to applicant** (CHMP/EMEA)

Clock stopped   Prepare response (Applicant) GMP/GCP/GLP
Inspections (CA)

Day 121   Response & revised SPC and labelling submitted (Applicant)   Sample testing

Response review (Rapporteurs)

Day150   Joint report on responses (Rapporteurs)

Comments/feedback from CHMP

Day 180   **List of outstanding issues or oral hearing** (CHMP)

Clock stopped   Oral hearing if necessary

Day 181   Final draft of English SPC, labelling and insert (Applicant)

Day 210   **Adoption of Opinion and Assessment Report** (CHMP)

Prepare to transfer opinion (EMEA)
Linguist review
Day 225   Opinion transferred to Commission (EMEA)   Product info (MS)

Prepare draft decision (Commission)

Day 229   Draft decision to Standing Committee + multilingual Annexes

Standing Committee consultation

Day 261   End of consultation

Prepare final decision

Day 277   **Decision signed by Director General on behalf of the Commissioner**

Figure 6.15 The Community marketing authorisation process.

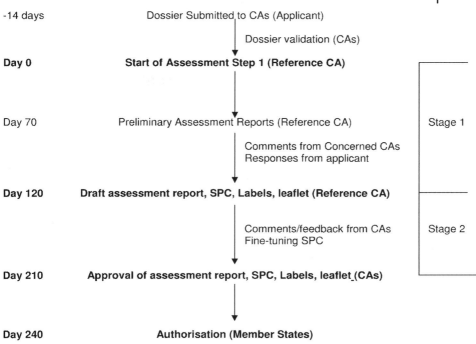

-14 days — Dossier Submitted to CAs (Applicant)

Dossier validation (CAs)

Day 0 — **Start of Assessment Step 1 (Reference CA)**

Day 70 — Preliminary Assessment Reports (Reference CA) — Stage 1

Comments from Concerned CAs
Responses from applicant

Day 120 — **Draft assessment report, SPC, Labels, leaflet (Reference CA)**

Comments/feedback from CAs — Stage 2
Fine-tuning SPC

Day 210 — **Approval of assessment report, SPC, Labels, leaflet (CAs)**

Day 240 — **Authorisation (Member States)**

Figure 6.16 Decentralised marketing authorisation process.

draft assessment report and the product information. During this phase, further fine-tuning of the SPC, leaflet and package insert may be required, depending on feedback from the Member States. At the end of 90 days the Member States must approve the assessment report, SPC, leaflet and package insert, after which they are allowed a further 30 days to enact national legislation granting the authorisations. If there is a failure to reach agreement between the Member States the issues can be referred to the EMEA/CHMP for arbitration. The procedure is outlined in Figure 6.16.

6.4.6
Mutual Recognition Procedure

An applicant must use the mutual recognition procedure to obtain marketing authorisations in other Member States, once an authorisation has been granted by a Member State. The applicant must submit the dossier to each of the concerned Member States where authorisation is requested, and identify the reference State that granted the original authorisation. Before starting the process the applicant is advised to ensure that the existing Member State dossier is fully up to date based on marketing experience. The applicant is also encouraged to engage in dialogue with the reference Member State on adjustments to the approved product information so as to ensure a smooth recognition procedure by the other Member States. The reference Member State has 90 days to update the original assessment report based on dossier updates and forward it to the concerned Member States together with the SPC, label and leaflet text that it has approved. Within 90 days of the receipt of these documents, the concerned Member States shall recognise the decision of the reference Member State and the approved summary of product characteristics, package leaflet and labelling by granting a marketing authorisation with a harmonised summary of product characteristics, package leaflet and labelling. During this phase, adjustments to the product information may be agreed, with the reference Member State acting as the coordinating interface with the applicant. If there is disagreement between Member States regarding the granting of an authorisation, the issue is referred to a coordinating group, consisting of Member States' representatives. If it cannot be resolved there, then it is referred to the CHMP for review and possible authorisation via the centralised procedure.

6.4.7
Plasma Master Files and Vaccine Antigen Master Files

Documentation providing detailed information on the characteristics, collection and control of human plasma used as the starting material for blood-derived products, or the production of antigens for vaccine products manufacture, can be submitted as stand-alone Plasma Master Files/Vaccine Antigen Master Files. This is intended to simplify the task of both the applicant and the authorities where the starting materials become constituents of many different products. Rather than reviewing the quality and safety of the starting materials as part of each individual marketing application, the files can be centrally submitted to the EMEA for evaluation and issue of certificates of compliance. A period of 90 days is allowed to compete the review process. The certificates can be used to support marketing applications through either centralised or national procedures. Certificates must be renewed annually.

6.5
Submission and Review Process in the US

Applications to market a chemical-based drug must be submitted to the Center for Drug Evaluation and Research (CDER) as a New Drug Application (NDA), whereas biologics are submitted to the CDER or Center for Biologics Evaluation and Research (CBER) as a Biologics Licensing Application (BLA). The fee for a full application has been set at $1 178 000 for the financial year 2008. If making a paper submission, the CDER require one compete archival copy, a number of review copies containing technical sections for the various reviewers, and a field copy of the quality section that must be sent directly to the local office. If using an electronic submission, only one copy is required as the FDA can load it on to their IT system for access by all concerned reviewers. It is a recommended standard practice for the applicant to request a pre submission meeting with the FDA, where procedural and other issues relating to the application can be discussed.

A designated project manager is used to coordinate the review; this person also acts as a primary point of contact with the applicant. Once received, an application is checked into a document control centre, where the archive copy is maintained. If in electronic format, the submission is loaded on to the IT system. Relevant sections of the dossier are then distributed to the review team and a preliminary technical screening is performed to check that the submission (including fees) is complete and that a worthwhile evaluation is possible. If deficiencies are found, a "refusal to file" letter must be sent to the applicant with 60 days, detailing the reasons why the submission is not acceptable. Otherwise, the application is filed and the formal review process begins. A flow chart outlining an NDA review is shown in Figure 6.17.

The review is conducted by a team of technical specialists drawn from different offices within the centres. They examine the dossier in great detail, which may involve a re-analysis or an extension of the analyses performed by the applicant. Physicians almost always conduct the medical review, and are responsible for evaluating the clinical sections of the submission. As this tends to be crucial to the assessment, they usually play a lead role in formulating an overall opinion of the product in terms of safety, efficacy and benefit-to-risk ratio. Other reviewers such as chemists, statisticians, pharmacologists, biopharmaceutical scientists and microbiologists (for anti-infective drugs) evaluate elements of the dossier specific to their areas of expertise. During the course of the review, the FDA will inform the applicant of easily correctable deficiencies. The applicant can submit an amendment to rectify such deficiencies immediately and thus allow the review to proceed smoothly. A 90-day conference between the applicant and the FDA may also be held for the purpose of discussing the overall progress of the review.

When they have completed their evaluations, the primary review team prepare reports that are then checked by their respective supervising officers. Parallel to the review, the project manager will ensure that any necessary GMP inspections of manufacturing facilities or testing of drug samples is completed by the relevant FDA offices. Finally, the project manager must coordinate the preparation of the recommendation resulting from the review process and the appropriate action letter, which will be used to officially communicate the outcome. These are then passed to the appropriate Director for final review and sign off of the action letter. In some cases, the Director may decide to obtain a second opinion from an Advisory Committee, made up of experts from outside the agency. A review by an Advisory Committee will involve public hearings where the applicants can present their case as to why the drug should be approved.

An action letter must be issued within 180 days of the start of the review, unless an extension has been agreed with the applicant. If the submission is fully acceptable an Approval Letter is issued. However, if there are problems remaining, the applicant will receive a Not Approvable Letter (major problems), or an Approvable Letter (minor problems) in the case of an NDA and a Complete Response Letter in the case of a BLA. These not only summarise the deficiencies that exist but also either outline what steps need to be taken to address the issues or explain why the application cannot be approved. A negative action letter will usually trigger an end-of-review meeting between the FDA and the applicant. This allows the applicant and FDA to discuss and

IND Review Process

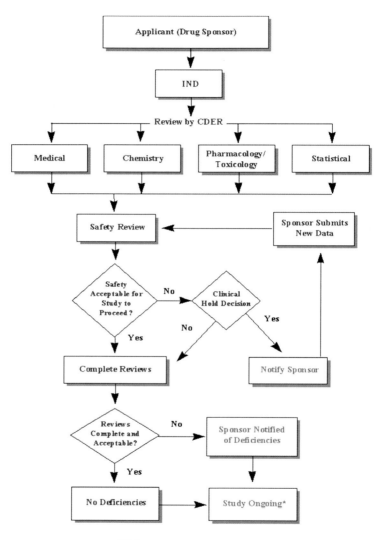

Figure 6.17 The NDA review process (from the CDER Handbook).

agree on what needs to be done in order to approve the application. Frequently, these may just focus on agreeing the final text of the labelling or other promotional material. A Summary Basis of Approval (SBA) document serves a similar purpose to the European Public Assessment Report (EPAR). It can be prepared by either the applicant or the FDA, and contains a summary of the safety and efficacy data on which the application was approved.

6.6
Chapter Review

This chapter examined the content and format of the marketing authorisation application as defined in the CTD. This is a harmonised format, agreed through the ICH process, which divides submissions into five modules, consisting of region-specific, summary, quality, non-clinical and clinical sections. The chapter then looked at the submission and review processes that exist in the EU and US. European authorisations may be granted using either a centralised community procedure, in which the EMEA are responsible for conducting the scientific evaluation, or via national procedures, in which the Competent Authority of a reference Member State takes a lead role in evaluating the application.

6.7
Further Reading

ICH Guidelines
  Common Technical Document (CTD), ICH M4, M4Q, M4S, M4E.
  Electronic Common Technical Document (eCTD) specifications.
  www.ich.org
European Regulations and Guidance
  Consolidated Human Medicines Directive 2001/83/EC, Title III.
  Regulation (EC) No. 726/2004, Title II, Chapter 1.
  The Rules Governing Medicinal Products in the European Union, Volume 2, Notice to Applicants, Medicinal Products for Human Use.
  http://ec.europa.eu/enterprise/pharmaceuticals/index_en.htm.
US Code of Federal Regulations & Guidance
  21 CFR Part 314, Subparts A, B, D, E & G.
  21 CFR Part 601, Subparts A & C.
  CDER Handbook.
  www.fda.gov

7
Authorisation of Veterinary Medicines

7.1
Chapter Introduction

The previous chapters examined the process for the development and authorisation of a drug product containing a new active ingredient for human use. This represents the most arduous path to market for any medicinal product. This chapter proceeds to examine the process of bringing a veterinary medicinal product to market. While the process shares most of the principles that apply to human drugs, there are some additional features that are unique to veterinary products. These include methods of use and the requirement to evaluate withdrawal periods and maximum residue limits in food-producing species.

7.2
Overview of Development Process for Veterinary Medicines

The process of developing and bringing to market veterinary medicines is fundamentally the same as that for human medicines. However, there are a number of features associated with veterinary medicines that can have either a positive or negative bearing on the overall process.

On the positive side, the fact that animals are the target species means that many of the ethical issues associated with the development of human medicines are avoided. At the outset, there is easy access to animal tissues for *in-vitro* investigations. Since all safety and efficacy trials are conducted in animals, the developer does not have to face the challenges of subject enrolment and ethical standards associated with human trials. The demonstration of efficacy in animals may also be easier. Studies can be conducted under controlled conditions, which avoid the compounding influences of diet or lifestyle that may be associated with human trials. The range of conditions of general interest to veterinary medicine tends to be narrower. The most common veterinary medicines are targeted at preventing or controlling infection, and are dominated by vaccines, antibiotics and anti-parasitic drugs. It is generally easier to establish the efficacy of such drugs where a defined effect is sought, in contrast to

Medical Product Regulatory Affairs. John J. Tobin and Gary Walsh
Copyright © 2008 WILEY-VCH Verlag GmbH & Co. KGaA, Weinheim
ISBN: 978-3-527-31877-3

many human conditions where only subtle responses may be experienced. The development of drug delivery systems may not require the same level of effort as for human medicines, as many target parasites are located either externally or in the gut. In addition, injection is an acceptable method of administration for vaccines or antibiotics.

On the down side, each target animal species will have to be individually investigated, as uniform responses may not be obtained among species. The developer must also address the concern that residues of chemical drugs in food-producing animals could pose a risk to human health. Thus, the metabolism and excretion of the drug must be thoroughly investigated with a view to establishing appropriate maximum residue limits and withdrawal periods that will assure the safety of food products derived from treated animals.

A further aspect that may add to the development work relates to the method of administration. In addition to the standards methods of administration associated with human medicines, veterinary medicines may be introduced to animals through medicated feeds. Such practices have been most commonly associated with the use of antibiotics or other substances to enhance animal performance under intensive farming systems. The use of medicated feeds has been phased out in Europe since the end of 2005, with the exception of coccidiostats and histomonostats intended to kill or inhibit protozoa. In the USA, however, medicated feeds are still commonplace. Thus, the developer may have to consider the stability and appropriate labelling of the drug when compounded into medicated feeds.

7.2.1
Pre-Clinical Studies

The pre-clinical phase of development of veterinary medical products follows a similar pattern to that for human drugs. The developer will focus on establishing the pharmaceutical and toxicological properties of the drug and generating data on the chemistry/identity, manufacture, control and stability of the product. Animal drugs, in general, have a greater probability of release to the environment than human drugs, through either animal excreta or, in the case of fish farming, directly into water. Thus, the developer is more likely to have to conduct specific ecotoxicology studies to assess the impact of the drug on the environment. In the case of chemical drugs intended for use in food-producing species, the non-clinical development must include the establishment and validation of appropriate analytical methods that enable the detection and quantification of residues of the drug or its metabolites in food produce such as meat, milk, eggs, or honey. Unlike human medicines, significant amounts of basic efficacy data may be generated in the pre-clinical phase of drug development by conducting studies in target species maintained under laboratory conditions. Immunological products will also require some specific types of study, particularly when prepared from live or modified organisms that could pose a threat in the wider environment. These would include investigations of the excretion of organisms from treated animals, the potential for re-emergence of virulence on passage through the host animals, and possible adverse impacts on nonvaccinated animals or other species. Generally, trials must be conducted in each target species under controlled bio-secure laboratory conditions before field trials can commence.

As with human pharmaceuticals, all safety studies must be conducted in accordance with the principles of Good Laboratory Practice (GLP). Similar to the International Conference on Harmonisation (ICH), the Veterinary International Conference on Harmonisation (VICH) has developed agreed guidelines on the types of test and other considerations that need to be addressed during the development and evaluation of veterinary medicines. A list of VICH guideline topics is provided in Table 7.1.

According to US regulations, drugs shipped to test facilities for non-clinical investigations must contain the following label warning:

> *Caution. Contains a new animal drug for investigational use only in laboratory research animals or for tests in vitro. Not for use in humans.*

7.2.2
Clinical Trials

Clinical trials must be conducted to establish safety in each target animal species which, in the majority of cases, will include food-producing species. Although the same types of animal may be involved in both pre-clinical and clinical trials, clear distinctions can be drawn between each type of study. Pre-clinical studies are only conducted in animals that are kept for the purposes of laboratory research, and that are usually maintained in a very controlled environment. Clinical trials, on the other hand, are conducted in animals that are representative of the normal conditions (field conditions) and purposes for which they are maintained.

The process of initiating animal clinical trials is less arduous than for human trials. The main concerns relate to ensuring that food from trial animals will not pose a risk to human health, and that the drug will not have an adverse effect on the environment. Of course, all trials should be conducted so as to minimise suffering and adverse impacts on animal welfare, but there is not the same level of ethical or safety concerns as applies to human trials. Studies may follow a similar pattern to human trials in that safety and efficacy may be established initially in clearly defined groups of animals. Investigations may then be expanded to explore the impact of different field conditions that could be encountered, such as climate, diet, or animal husbandry practices. In food-producing species, particular attention must be given to residue depletion and elimination studies. The potential to develop resistance must also be investigated in the case of antibiotic and anti-parasitic drugs.

7.2.3
Good Clinical Practices

VICH guideline GL9 sets out the principles of Good Clinical Practices for clinical trials of veterinary medicines. The guideline has not been incorporated into specific

Table 7.1 Topics addressed by VICH guidelines.

Document no.	Topic	Title of guidelines
GL1	Validation definitions	Validation of analytical procedures: definition and terminology
GL2	Validation methodology	Validation of analytical procedures: methodology
GL3	Stability 1	Stability testing of new drug substances and products
GL4	Stability 2	Stability testing for new dosage forms
GL5	Stability 3	Stability testing: photostability testing of new drug substances and products
GL6	Ecotox Phase I	Environmental impact assessments (EIAs) for veterinary medicinal product (VMPs) Phase 1
GL7	Anthelmintics General	Efficacy of anthelmintics: general requirements
GL8	Stability premixes	Stability testing for medicated premixes
GL9	GCP	Good Clinical Practices
GL10	Impurities substances	Impurities in new veterinary drug substances
GL11	Impurities products	Impurities in new veterinary medicinal products
GL 12	Anthelmintics bovine	Efficacy of anthelmintics: specific recommendations for bovines
GL13	Anthelmintics ovine	Efficacy of anthelmintics: specific recommendations for ovines
GL14	Anthelmintics caprine	Efficacy of anthelmintics: specific recommendations for caprines
GL15	Anthelmintics equine	Efficacy of anthelmintics: specific recommendations for equines
GL16	Anthelmintics porcine	Efficacy of anthelmintics: specific recommendations for porcines
GL17	Stability: Biotechnologicals/ biologicals	Stability testing of new biotechnological/biological products
GL18	Impurities: Residual Solvents	Impurities: residual solvents in new veterinary medicinal products, active substances and excipients
GL19	Anthelmintics canine	Efficacy of anthelmintics: specific recommendations for canine
GL20	Anthelmintics feline	Efficacy of anthelmintics: specific recommendations for feline
GL21	Anthelmintics poultry	Efficacy of anthelmintics: specific recommendations for poultry
GL22	Safety reproduction	Studies to evaluate the safety of residues of veterinary drug in human food: reproduction studies
GL23	Safety genotoxicity	Studies to evaluate the safety of residues of veterinary drug in human food: genotoxicity testing
GL24	Pharmacovigilance	Pharmacovigilance of veterinary medicinal products: management of Adverse Event Reports (AERs)

Table 7.1 (*Continued*)

Document no.	Topic	Title of guidelines
GL25	Biologicals	Testing of residual formaldehyde
GL26	Biologicals	Testing of residual moisture
GL27	Antimicrobial Resistance	Guidance on pre-approval information for registration of new veterinary medicinal products for food producing animals with respect to antimicrobial resistance
GL28	Safety carcinogenicity	Studies to evaluate the safety of residues of veterinary drug in human food: carcinogenicity testing
GL29	Pharmacovigilance PSU	Pharmacovigilance of Veterinary Medicinal Products: Management of Periodic Summary Update Reports (PSUs)
GL30	Pharmacovigilance list of terms	Pharmacovigilance of Veterinary Medicinal Products: controlled list of terms
GL31	Safety: Repeat-dose toxicity test	Studies to evaluate the safety of residues of veterinary drugs in human food: Repeat-dose toxicity testing
GL32	Safety: Developmental toxicity test	Studies to evaluate the safety of residues of veterinary drugs in human food: Developmental toxicity testing
GL33	Safety: General approach to testing	Studies to evaluate the safety of residues of veterinary drugs in human food: General approach to testing
GL34	Biologicals: Mycoplasma	Test for the detection of Mycoplasma contamination
GL35	Pharmacovigilance: ESTD	Pharmacovigilance: Electronic Standards for Transfer of Data
GL36	Safety: microbiological ADI	Studies to evaluate the safety of residues of veterinary drugs in human food: General Approach to establish a microbiological ADI
GL37	Safety: repeat dose chronic toxicity	Studies to evaluate the safety of residues of veterinary drugs in human food: Repeat-dose Chronic Toxicity Testing
GL38	Ecotoxicity Phase II	Environmental Impact Assessment (EIAs) for Veterinary Medicinal Products (VMPs) – Phase II
GL39	Quality: specifications	Test Procedures and Acceptance Criteria for new Veterinary Drug Substances and New medicinal Products: Chemical Substances
GL40	Quality: specifications	Test Procedures and Acceptance Criteria for new Biotechnological/Biological Veterinary Medicinal Products
GL41	TAS: reversion to virulence	Examination of live Veterinary Vaccines in Target Animals for Absence of Reversion to Virulence
GL42	Pharmacovigilance: Data elements	Pharmacovigilance: Data Elements for Submission of Adverse Events Reports

regulations, but it should be followed as it provides assurance to both the regulatory authorities and the public at large as to the following:

- That the data and reported results are complete, correct and accurate,
- That the welfare of the study animals and the safety of the study personnel involved in the study are ensured.
- That the environment and the human and animal food chains are protected.

As with human GCP, the guideline addresses:

- the duties and requirements of sponsors and investigators;
- the preparation of trial protocols and supporting standard operating procedures;
- the recording, storage and reporting of data;
- the notification of adverse events; and
- the monitoring and quality assurance auditing of the trial.

Unlike human trials, Institutional Review Boards/Independent Ethics Committees are not involved, while informed consent is only required from the owner of the trial animals. In addition to the standard items associated with human trials, aspects such as management and housing of animals, diet and disposal of trial animals and their produce should be included in the trial protocol. Studies may be blinded from the investigators in order to avoid bias in the reporting of animal observations.

7.3
Authorisation of Clinical Trials in the EU

Unlike human trials, animal trials have not been specifically regulated by Directive at EU level. Instead, it is left up to Member States to adopt national legislation to protect food safety and the environment. In Ireland, trials are regulated via regulation 19 of The Animal Remedies Regulations 2005 (S.I. No. 734 of 2005.). This requires that a trial site obtain a "research licence" from the Minister of Agriculture before administering non-authorised drugs to animals. The Minister will consult with the Irish Medicines Board before issuing a licence. The Minister must also authorise an appropriate withdrawal period before produce from trial animals can enter the food chain. These may be based on reference withdrawal periods from Directive 2001/82/EC (see Table 7.2), and augmented by additional safety factors as appropriate.

Table 7.2 Reference minimum withdrawal periods from Directive 2001/82/EC.

Produce	Withdrawal period
Eggs	7 days
Milk	7 days
Meat from poultry or mammals	28 days
Fish	500 degree–days[a]

[a]Degree–days are calculated by multiplying the water temperature by the number of days.

7.4
Authorisation of Clinical Trials in the US

The US regulates animal trials mainly through the requirement to inform the authorities before non-approved veterinary pharmaceuticals are shipped to trial sites. Depending on whether it is classified as a drug or a biologic, information must be submitted to either the FDA Center for Veterinary Medicines (CVM) or the United States Department of Agriculture (USDA) Center for Veterinary Biologics (CVB).

Before shipping an animal drug, a "Notice of Claimed Investigational Exemption for a New Animal Drug" must be submitted to the CVM on FDA Form 3458, accompanied by the information outlined in Figure 7.1. The specific regulations are contained in 21 CFR Part 511, New Animal Drugs for Investigational Use. The investigational drug product must be labelled with the following warning:

> *Caution. Contains a new animal drug for use only in investigational animals in clinical trials. Not for use in humans. Edible products of investigational animals are not to be used for food unless authorization has been granted by the U.S. Food and Drug Administration or by the U.S. Department of Agriculture.*

If intended for food-producing animals, the labels should also bear the statement:

> *No official withdrawal time has been established for this product under the proposed investigational use.*

Produce from trial animals may not enter the food chain unless authorised by the CVM on the basis of data showing that residues will be either safe or not present in produce from animals treated at the maximum dosage with the minimum allowed withdrawal period. The CVM must be notified of the date and place of slaughter at least 0 days prior to each shipment for slaughter. The regulations also address the retention of records and the competence of study investigators. Records of drug shipments must be maintained for at least 2 years after the date of shipment. Data and results of trials must be retained for either 2 years after completion of the trial, or

1. The identity of the new animal drug
2. Labelling and other pertinent information supplied to investigators (study protocols, non-clinical laboratory data, etc.).
3. The name and address of each clinical investigator.
4. Approximate number of animals to be treated or quantities of drug to be shipped.
5. If food-producing species, a commitment that produce will not be used without prior authorisation, the approximate start and finish date of the trials, and information on the dosage and animals.
6. Information on Contract Research Organisations, if involved
7. Environmental assessment or exemption declaration

Figure 7.1 Information supplied to the CVM in support of a Notice of Claimed Investigational Exemption (NCIE) for an Investigational New Animal Drug (INAD).

1. Permit or letter of authorisation issued by the state where each trial will be conducted
2. Tentative list of proposed recipients
3. Description of product and available preliminary information on safety and efficacy
4. Copies of labels
5. Study protocol
6. Data showing that the product is unlikely to adversely affect food produce from the animals
7. Commitment from investigator to provide additional information before movement of food-producing animals, if applicable
8. Information on possible environmental impact

Figure 7.2 Information supplied to the CVB in support of a request for authorisation to ship an experimental biologic product.

approval of the drug, whichever is relevant. The sponsor must ensure that the investigator is suitably qualified and experienced to conduct the studies.

Regulations covering experimental biologic products are contained in 9 CFR Part 103 Experimental Production, Distribution and Evaluation of Biological Products Prior to Licensing. In this case, authorisation must be obtained from the CVB before the product is shipped. Requests for authorisation must be supported by the information outlined in Figure 7.2. Experimental biologics should not be manufactured in the same facility as licensed biologics, unless authorised by the authorities on the basis that adequate measures are in place to prevent contamination of licensed products. Product labels must bear the statement:

Notice! For Experimental Use Only – Not For Sale.

Animals may not be moved for 14 days after administration of the experimental product, and records on the disposition of trial animals must be retained for 2 years. Bio-security issues will be of particular concern to the environmental impact assessment where trials involve live organisms or genetically modified organisms, either in vaccine challenge studies or as experimental products.

7.5
Maximum Residue Limits

A unique feature of the development of veterinary medicines for food-producing species is the need to consider the establishment of Maximum Residue Limits (MRLs) for the drug or its metabolites in food produce. A MRL is defined in the European Union as:

The maximum concentration of residue resulting from the use of a veterinary medicinal product (expressed in mg/kg or μg/kg on a fresh weight basis) which may be accepted by the Union to be legally permitted or recognised as acceptable in or on a food.

The requirement to establish MRLs stems from concerns that the residues of veterinary pharmaceuticals may have undesirable toxicological or pharmaceutical effects in humans, if present in food. With antibiotics in particular, it is known that residues can present a risk to human health by favouring the emergence of antibiotic resistance in bacteria of the human gut flora, or by disrupting the normal healthy intestinal flora that provide a colonisation barrier against less-desirable organisms. MRLs are generally not considered necessary for biopharmaceutical substances such as vaccines, although pharmacologically active excipients, adjuvants, or preservatives used in such products, will need to be assessed.

MRLs are derived using a two-stage process. The first step is to establish an acceptable daily intake (ADI), which represents the quantity of substance and/or its residues, expressed as μg or mg per kg bodyweight, that can be ingested daily over a lifetime without any appreciable health risk to exposed individuals. This will be based on a review of the various pharmacological, toxicological, microbiological and other tests undertaken to demonstrate the safety of the substance. This usually permits identification of the no observed (adverse) effect level (NO(A)EL) or, in certain cases the lowest observed (adverse) effect level (LO(A)EL), with respect to the most sensitive parameter in the most sensitive appropriate test species or, in some cases, in humans. An uncertainty factor or safety factor is then applied to take account of the inherent difficulties in extrapolating animal toxicity data to humans and the variations between humans.

Once an ADI has been established, it can be used to estimate MRLs for different food commodities which provide assurance that the likely dietary intake of residues will remain below this threshold. Typically, calculations are based on an arbitrary average human body weight of 60 kg and an assumed average daily consumption of 500 g of meat (made up of 300 g muscle, 100 g liver, 50 g kidney and 50 g fat) together with 1.5 l of milk and 100 g of eggs or egg products. Consideration must also be given to residue depletion patterns and the possibility of persistence in specific organs when proposing MRLs. The process should result in the proposal of MRLs for all the following food commodities from all species intended for human consumption; meat of mammals and poultry (muscle, fat (fat and skin where appropriate), liver and kidney), meat of fin fish (muscle and skin in natural proportions), milk, eggs and honey.

At international level, a Joint Expert Committee on Food Additives (JECFA) under the auspices of the United Nations Food & Agriculture Organisation (FAO) and the World Health Organization (WHO) have undertaken safety evaluations of residues of veterinary medicines among other contaminants in food produce. As a result, MRLs for a number of common veterinary drugs have been published in the "Codex Alimentarius". As these have been developed in the interests of safe food and fair trade, World Trade Organisation and other international agreements promote the use of such limits by national authorities, unless there is strong scientific justification to the contrary. For new drugs, proposals as regards residues will have to be addressed as part of the authorisation process.

7.6
Authorisation of Veterinary Medicines in the EU

The authorisation of veterinary medicines containing a new chemical entity for use in food-producing animals involves a two-stage application process: (i) an application to establish MRLs; and (ii) an application for a marketing authorisation.

7.6.1
Applications to Establish MRLs

The Community procedure for establishing MRLs of veterinary medicines in foodstuffs of animal origin is set down in Council Regulation (EEC) No. 2377/90 as amended by Council Regulation (EC) No. 1308/1999. The regulations are supported by detailed guidance contained in The Rules Governing Medicinal Products in the European Union, Volume 8 - Maximum Residue Limits.

Applications to establish MRLs for new pharmacologically active substances must be submitted to the European Medicines Agency (EMEA) at least 6 months in advance of an application for a marketing authorisation. In order to avoid delays, manufacturers are advised to submit an application once all the necessary data are available, as a product authorisation cannot be granted unless established MRLs are in place. The EMEA should be notified of the intent to submit an application 3 to 4 months in advance of the anticipated date, so that a Rapporteur and Co-Rapporteur can be appointed from among the members of the Committee for Veterinary Medicinal Products (CVMP).

Applications must be submitted using the form shown in Figure 7.3, and supported by a technical dossier consisting of a safety file and a residue file. The contents of the dossier are specified in Annex V to Regulation 2377/90 on maximum residue limits, and outlined in Figure 7.4. The safety file focuses on providing information on pharmacological and toxicological studies that leads to a proposed ADI for the substance or its metabolites. The residue file provides data to indicate how the residues will be distributed and depleted, together with proposed MRLs and validated analytical methods for detection of the residues in food commodities. As with marketing applications, signed expert reports that provide a critical assessment and commentary on the data form a fundamental part of the files.

7.6.2
Review of Applications and Establishment of MRLs

The review of MRL applications is similar to that for centralised marketing authorisations conducted by the EMEA, in that rapporteurs are responsible for the hands-on evaluation, which is then reported back to the CVMP for consideration. If there are outstanding issues, a list of questions is forwarded to the applicant for his or her response. Otherwise, a formal opinion is prepared and presented to the Commission for legal implementation as a decision. The maximum time allowed to deliver a CVMP opinion is 120 days, excluding the time taken for an initial validation of the application and review of responses form the applicant where necessary.

APPLICATION FORM

APPLICATION FOR THE ESTABLISHMENT OF MRL(s) FOR AN ACTIVE SUBSTANCE TO BE USED IN VETERINARY MEDICINAL PRODUCTS IN ACCORDANCE WITH COUNCIL REGULATION (EEC) No. 2377/90 AS AMENDED

PART I: Administrative Data

Rapporteur:
Co-rapporteur:
Name of Substance for review, using INN (where attributed):
Name and address of applicant:
Please summarise the anticipated pattern of veterinary use: Target Species   Major indications   Dose regimen
Is the substance used in veterinary medicinal products as - active ingredient - excipient, preservative, etc?
Name, address, telephone number and fax number of company contact point for all correspondence arising in connection with the application:

Figure 7.3 MRL application form.

PART II: **SUMMARY OF THE EVALUATION PROPOSED BY THE APPLICANT**

Name of substance for review:
Is this application for inclusion in: - Annex I   (MRL values) - Annex II   (no MRLs necessary)
Indicate the most relevant NOEL for the evaluation of the safety of residues (mg/kg bw/day):
Reference to relevant study (including location in the dossier):
Uncertainty factor proposed:
ADI proposed (mg/kg bw):
ADI x 60 =   (acceptable daily intake per adult; mg)
MRLs proposed (µg/kg) MRL   Food Commodity   Marker Residue
Method of analysis proposed for residue monitoring Method: Limit of quantification: Reference (including location in the dossier):

I hereby certify that all information relating to the establishment of MRLs for the above-mentioned substance, whether favourable or unfavourable, has been submitted with this application.

Date   Signature

Figure 7.3 (*Continued*)

A. Safety file

A.0. Expert Report

A.1. Precise identification of the substance concerned by the application
Names (INN, IUPAC, CAS) Therapeutic & Pharmacological class, Structural & molecular formulae, impurities, Description of physical properties

A.2. Pharmacology
A summary of the results of the various pharmacological studies, with particular emphasis on results which may be relevant to the evaluation of the safety of residues of the substance.
      A.2.1 Pharmacodynamics
      A.2.2 Pharmacokinetics

A.3. Toxicological studies
      3.1 Single-dose toxicity
      3.2 Repeated-dose toxicity
      3.3 Tolerance in the target species of animal
      3.4 Reproductive toxicity, including teratogenicity
          3.4.1 Study of the effects on reproduction
          3.4.2 Embryotoxicity/foetotoxicity, including teratogenicity
      3.5 Mutagenicity
      3.6 Carcinogenicity

A.4. Studies of other effects
      4.1 Immunotoxicity
      4.2 Neurotoxicity
      4.3 Microbiological properties of residues
      4.3.1 Potential effects on the human gut flora
      4.3.2 Potential effects on the micro-organisms used for industrial food processing
      4.4 Observations in humans

A5 Safety evaluation of residues
      5.1 Proposal for an acceptable daily intake (ADI)
      5.2 Alternative limits
      5.3 When no limit can be set

B. Residue File

B.0. Expert report

B.1. Precise identification of the substance concerned by the application
The substance concerned should be identified in accordance with point A.1. However, where the application relates to one or more veterinary medicinal products, the product itself should be identified in detail.

B.2. Residue studies
      2.1 Pharmacokinetics (absorption, distribution, metabolism, excretion)
      2.2 Depletion of residues
      2.3 Elaboration of MRLs

B.3. Analytical method for the detection of residues
      3.1 Description of the method
      3.2 Validation of the method
      (Specificity, Accuracy, Precision, Limit of detection, Limit of quantification, Practicability and applicability under normal laboratory conditions, Susceptibility to interference, Stability)

Figure 7.4 Outline of the MRL dossier.

Table 7.3 Possible outcomes of a MRL application.

Annex I	Final MRLs established
Annex II	No specific residue limits required for protection of human health
Annex III	Provisional MRLs valid for a maximum of 5 years
Annex IV	No safe residue limits can be established. Substance may not be administered to food-producing animals

The outcome of the process will result in the inclusion of the substance in one of four Annexes to the regulation. If the conclusion is that safe MRLs can be established, then final MRLs for the substance are added to Annex I. A substance will be inserted in Annex II if it is concluded that residues cannot cause a risk to human health. If some uncertainty remains over the proposed MRLs, then these can be included in Annex III as provisional MRLs for a maximum period of 5 years. Finally, if it is concluded that no safe residue limits can be set, the substance is placed in Annex IV, which means that it cannot be used in medicines for treatment of food-producing animals (Table 7.3).

7.6.3
Marketing Authorisations

The procedures and timelines for obtaining marketing authorisations for veterinary products are essentially the same as those for human medicines. Extensive guidance can be found in The Rules Governing Medicinal Products in the European Union, Volume 6 – Notice to Applicants, Veterinary Medicinal Products. The products that may follow centralised procedures and receive Community authorisations are shown in Figure 7.5. All other products must obtain authorisations from national

Mandatory

- Biotech products based on recombinant DNA, gene expression or hybridoma/monoclonal antibody technologies.
- Products intended primarily for use as performance enhancers in order to promote the growth of treated animals or to increase yields from treated animals.

Optional

- Other new active substances.
- Medicinal products that contribute significant therapeutic, scientific or technical innovation, or are in the interests of animal health.
- A generic copy of a centrally authorised product.
- Immunological products for the treatment of animal diseases that are subject to Community prophylactic measures.

Figure 7.5 Veterinary products that may be authorised via centralised procedures.

Competent Authorities using Mutual Recognition, Decentralised or National procedures. In the case of Community authorisations, the CVMP and the Standing Committee for Veterinary Medicinal Products perform roles identical to those played for human medicines by the Committee for Medicinal Products for Human Use (CHMP) and the Standing Committee for Human Medicinal Products. As already noted, authorisations cannot be granted for use of veterinary products in food-producing animals unless all active substances are entered in Annexes I, II or III of the MRL regulations. Authorisations of vaccines may be limited to certain states or regions, on the grounds that immunisation would interfere with diagnostic procedures used to prevent, control or eradicate the disease in other areas. For example, Ireland as an island nation would tend to pursue policies of testing to eradicate and achieve disease-free status in favour of vaccination strategies.

7.6.4
Presentation of the Dossier

The requirements for the contents of an application dossier are set out in Annex I of Directive 2001/82/EC. The dossier should be assembled in four parts, as outlined in Figure 7.6. The application form, Summary of Product Characteristics (SPC) and labelling, expert reports and quality sections are quite similar in content to human

PART 1   Administrative Documentation and Summary of the Dossier
A.   Administrative Data (Application form, etc.)
B.   Summary of Product Characteristics Label and Package Insert
C.   Expert Reports (Quality, Safety, Efficacy)

PART 2   Quality Documentation
A.   Qualitative and Quantitative Composition
B.   Description of Manufacturing Methods
C.   Control of Starting Materials
D.   Specific measures concerning prevention of transmission of animal Spongiform Encephalopathies (TSEs) (if applicable)
E.   Control Testing of Intermediates
F.   Control Testing of Finished Product
G.   Stability Tests
H.   GMO Risk Assessment (if applicable)

PART 3   Safety and Residues Documentation
A.   Safety Testing (Pharmacology, Toxicology, Ecotoxicity)
B.   Residue Testing (Metabolism and residue kinetics, Routine analytical method for the detection of residues)

PART 4   Efficacy Documentation
Pre-clinical and Clinical Data
A.   Pharmacology (Pharmacodynamics, Pharmacokinetics)
B.   Tolerance in the Target Species
C.   Resistance

Figure 7.6 Outline of the dossier structure for a marketing application for a veterinary product.

medicines submissions. The target species and relevant withdrawal periods for food-producing animals must be stated in the SPC and labelling. The labels should also bear the statement "For animal treatment only" and, where applicable, ". . . to be supplied only on veterinary prescription". Data in Parts 3 and 4 are presented on the basis of the purpose for which the study was conducted – that is, for safety or efficacy. This differs from human medicines dossiers, where the data are segregated on the basis of how they were obtained (i.e. non-clinical or clinical studies). The type of information presented in the Safety and Efficacy Parts of the dossier will vary, depending on whether it is for an immunological product, or not. For example, residues or resistance documentation would not normally form part of a vaccine dossier.

7.7
Approval of Veterinary Medicines in the US

Depending on whether it is classified as a drug or a biologic, the applicant must submit either a New Animal Drug Application (NADA) to the FDA CVM or veterinary biologics licence applications to the USDA CVB.

7.7.1
New Animal Drug Application

The requirements for a NADA are set out in 21 CFR 514.1 and summarised in Figure 7.7. A FDA-356v form must accompany each submission (see Figure 7.8). This is similar to the form used for human drug submissions in that it captures basic information on the submission and provides a checklist for accompanying documents or materials. The labelling should include instructions on how to use the drug in the manufacture of medicated feeds, where such use is intended. The CVM may request samples of the drug product itself, reference materials or standards, and samples of food produce from treated animals in order that they can, in conjunction

(1) Identification.
(2) Table of contents and summary.
(3) Labeling.
(4) Components and composition.
(5) Manufacturing methods, facilities, and controls.
(6) Samples (on request)
(7) Analytical methods for residues.
   (also include proposed tolerance or withdrawal period and supporting data)
(8) Evidence to establish safety and effectiveness.
(9) Statement on GLP compliance status of laboratory studies
(10) Environmental assessment.
(11) Freedom of Information Summary (for public disclosure)
(12) Veterinary feed directive (if applicable).
(13) Patent information

Figure 7.7 Outline of requirements for a NADA submission.

DEPARTMENT OF HEALTH AND HUMAN SERVICES FOOD AND DRUG ADMINISTRATION	**NEW ANIMAL DRUG APPLICATION** (Drugs for Animal Use) (Title 21, CFR 514)	Form Approved: OMB No. 0910-0032 Expiration Date: December 31, 2007
		NADA

DRUG PRODUCT

ESTABLISHED NAME *(e.g. USP/USAN)*		PROPRIETARY NAME

DOSAGE FORM	PROPOSED INDICATIONS FOR USE	SPECIES

PROPOSED MARKETING STATUS *(Check one)*: ☐ PRESCRIPTION PRODUCT (Rx)   ☐ OVER-THE-COUNTER PRODUCT (OTC)

NAME OF APPLICANT	ADDRESS *(Street Number, City, and ZIP Code)*

NOTE: No application may be filed unless a completed application form has been received.

☐ ORIGINAL APPLICATION (21 CFR 514.1(a))

☐ ABBREVIATED ORIGINAL APPLICATION

☐ AMENDMENT TO AN UNAPPROVED ORIGINAL APPLICATION *(21 CFR 514.6)*

☐ SUPPLEMENT TO AN APPROVED APPLICATION (21 CFR 514.8(a))

☐ AMENDMENT TO AN UNAPPROVED SUPPLEMENT TO AN APPROVED APPLICATION *(21 CFR 514.6)*

☐ SPECIAL SUPPLEMENT TO AN APPROVED APPLICATION --CHANGES BEING EFFECTED *(21 CFR 514.8(e))*

REASON FOR SUBMISSION:

Paperwork Reduction Act Statement

A federal agency may not conduct or sponsor, and a person is not required to respond to, a collection of information unless it displays a currently valid OMB control number. Public reporting burden for this collection of information averages 242.6 hours per response, including time for reviewing instructions, searching existing data sources, gathering and maintaining the necessary data, and completing and reviewing the collection of information. Send comments regarding the burden estimate or any other aspect of this collection of information to: Food and Drug Administration
Center for Veterinary Medicine (HFV-12)
Attn.: Assistant Records Control Officer
7500 Standish Place
Rockville, MD 20855

INSTRUCTIONS FOR PREPARING AND SUBMITTING THE NEW ANIMAL DRUG APPLICATION

i.   Prepare three identical copies of the submission.

ii. Identify front cover of each copy with the name of the applicant, the proprietary name (if available), the name of the new animal drug and the dosage form.

iii. Use separate pages for each numbered heading consistent with the sub-paragraphs of this application form. (See reverse side of this page).

iv. Number the pages of the new animal drug application. Each copy should bear the same page numbering.

v.   Submit separate applications for each different dosage form of the proposed drug.

vi. Basic information pertinent to a dosage form should be made by reference to volume and page of the application containing such information. Include in each application information applicable to the specific dosage form, such as, labeling, composition, stability data, efficacy data, method of manufacture and references to appropriate investigational new animal drug applications and master file(s).

vii. Submit applications to: Food and Drug Administration
Center for Veterinary Medicine (HFV-199)
7500 Standish Place
Rockville, MD 20855

viii. Prepare amendments, supplements, reports and other correspondence in the above format. Identify the submission with the assigned NADA number.

ix. If the submission is a supplemental application, full information shall be provided on each proposed change concerning any statement made in the approved application.

x.   Submit page 1 and 2 of this form with each submission.

FOR FDA USE ONLY

FORM FDA 356V (12/04)   **Page 1**   PSC Media Arts (301) 443-1090   EF

Figure 7.8 FDA form 356v.

The following information in the cited section of 21 CFR 514 shall constitute the requirements of this application. Please check the information submitted.

> **NOTE:** 1. An original application shall include all of the following sections.
> 2. A supplement or amendment shall include only those sections necessary.

☐ 1. **IDENTIFICATION** 21 CFR 514.1(b)(1)

☐ 2. **TABLE OF CONTENTS AND SUMMARY** 21 CFR 514.1(b)(2)
A table of contents and summary of information to describe the chemistry of the proposed drug and product, the clinical purpose and a summary of laboratory and clinical studies.

☐ 3. **LABELING** 21 CFR 514.1(b)(3)
Copies of each proposed label.

☐ 4. **COMPONENTS AND COMPOSITION** 21 CFR 514.1(b)(4)
A list of all articles used as components of the drug product.
A statement of composition of the drug product.
A complete description of the fermentation of antibiotic drug substances.

☐ 5. **MANUFACTURING METHODS, FACILITIES, AND CONTROLS** 21 CFR 514.1(b)(5)
A detailed description of the manufacturer, personnel, facilities/equipment, new animal drug substance synthesis, raw material controls (specifications, tests and methods), manufacturing instructions, finished product analytical controls (specifications, tests and methods), stability program, container/packaging, and lot control number system.

☐ 6. **SAMPLES** 21 CFR 514.1(b)(6)
Samples to be submitted only upon the Center's request.

☐ 7. **ANALYTICAL METHODS FOR RESIDUES** 21 CFR 514.1(b)(7)
Method(s) and data to enable determination of residues of the drug in food-producing animals.

☐ 8. **EVIDENCE TO ESTABLISH SAFETY AND EFFECTIVENESS** 21 CFR 514.1(b)(8)
Data/information to permit evaluation of the safety and effectiveness of the drug product for the claim(s) proposed in the proposed species.

☐ 9. **GOOD LABORATORY PRACTICE COMPLIANCE** 21 CFR 514.1(b)(12)(iii)
A statement of compliance or non-compliance to good laboratory practices (21 CFR 58) of each nonclinical laboratory study.

☐ 10. **ENVIRONMENTAL ASSESSMENT** 21 CFR 514.1(b)(14)
An environmental assessment (21 CFR 25.40) containing data/informaiton to permit evaluation of the environmental safety of the drug product or a claim for a categorical exclusion from preparing an environmental assessment (21 CFR 25.33), as appropriate.

☐ 11. **FREEDOM OF INFORMATION SUMMARY** 21 CFR 514.11
A summary prepared according to Agency guidelines.

☐ 12. **OTHER** *(Specify)*

The undersigned official submits this application for a new animal drug pursuant to section 512(b) of the Federal Food, Drug, and Cosmetic Act. It is understood that the labeling and advertising for the new animal drug will prescribe, recommend, or suggest its use only under the conditions stated in the labeling which is part of this application and if the article is a prescription new animal drug, it is understood that any labeling which furnishes or purports to furnish information for use or which prescribes, recommends, or suggests a dosage for use of the new animal drug will also contain, in the same language and emphasis, information for its use including indications, effects, dosages routes, methods, frequency, and duration of administration, any relevant hazards, contraindications, side effects, and precautions contained in the labeling which is part of this application in accordance with 21 CFR 201.105. It is understood that all representations in this application apply to the drug produced until changes are made in conformity with 21 CFR 514.8. It is further understood that new animal drugs as defined in 21 CFR 510.3, intended for use in the manufacture of animal feeds in any State will be shipped only to persons who may receive such drugs in accordance with 21 CFR 510.7. Furthermore, the applicant certifies that the services of any person debarred under FFDCA, subsection 306(a) or (b), have not and will not be used in any capacity with this application.

> NOTE: This application must be signed by the applicant or by an authorized attorney, agent, or official. If the applicant does not have a place of business within the United States, the application must also provide the name and address of and be countersigned by an authorized agent or official residing or maintaining a place of business within the United States.

(WARNING: A willfully false statement is a criminal offense. U.S.C. Title 18, sec. 1001.)

SIGNATURE OF RESPONSIBLE OFFICIAL OR AUTHORIZED AGENT	TITLE OF AUTHORITY	DATE

FORM FDA 356V (12/04) **Page 2**

Figure 7.8 *(Continued)*

with the USDA Food Safety and Inspection Service (FSIS), establish and verify the analytical procedures for the drug or its residues. Unlike in Europe, the applicant does not have to submit a separate residues application, but instead can include proposed tolerances and withdrawal periods in the residue section of the NADA application. The FDA will then set final tolerances and withdrawal periods as part of the conditions for approval of the new drug. Where they exist, the FDA may rely on international residue limits established by the Codex Alimentarius Commission. The FDA can approve a new drug as prescription-only, non-prescription, or as a drug for administration in animal feed under the direction of a veterinarian – a Veterinary Feed Directive (VFD) drug. A VDF categorisation allows general access to drugs that may enhance animal health or performance, while at the same time reducing the risk of abuse or drug resistance that might exist, if there was unrestricted availability. The timelines and process for review and approval of a new animal drug are essentially the same as those for a human drug.

7.7.2
Approval of Veterinary Biological Products

In order to market a new veterinary biological product, a manufacturer must submit separate licence applications for the product and the establishment to the CVB. This differs from human biological products, where product and the facilities information may be combined in one Biologics License Application (BLA) submission. The requirements for each licence application are set out in 9 CFR Part 102. The information required for a Veterinary Biologics Establishment License application is outlined in Figure 7.9. A prime concern of establishment licensing is to ensure that viruses or toxins can be handled safely under appropriate bio-containment conditions. Thus, evidence must be provided as regards the suitability of the facilities and the competence of the operating staff. The product license application focuses on establishing the safety and efficacy of the biologic product. As bio-safety is also a critical aspect of many biologic products, submissions will include data, as relevant, on the purity of master seeds/cell stocks, the stability of genetically modified organisms, the validation of inactivation procedures for killed organisms, and the purity of products derived from organisms. The type of information required for approval of a Veterinary Product Licence Application is outlined in Figure 7.10.

1. Application Form (APHIS Form 2001)
2. Articles of incorporation
3. Water quality statement from local authorities certifying compliance with effluent waste regulations
4. Application for at least one Product License supported by an outline of production, information on master seeds/master cells and bio-containment if applicable, and protocols for studies of host animal safety and efficacy
5. Qualifications of personnel operating the facility

Figure 7.9 Outline of requirements for a Veterinary Biologics Establishment License.

1. Application Form (APHIS Form 2003)
2. Outline of production
3. Test reports and research data sufficient to establish purity, safety, potency, and efficacy of the product
4. Legends designating facilities used for each step
5. Labelling

Figure 7.10 Outline of requirements for a Veterinary Biologics Product Licence Application.

7.8
Chapter Review

This chapter examined the process for bringing veterinary medicinal products to market. This is quite similar to the process for human drugs, in that the developer must progress through pre-clinical, clinical and marketing authorisation phases. However, some differences of note do exist. As clinical trials do not involve human subjects, but rather animals maintained under the normal field conditions and purposes for which they are reared, they do not come under the same degree of regulatory oversight. The establishment of MRLs and appropriate withdrawal periods is an additional step that must be completed in the case of chemical-based veterinary pharmaceuticals intended for use in food-producing species. The USDA, through its Centre for Veterinary Biologics, rather than the FDA, is responsible for the oversight and approval of veterinary biologics.

7.9
Further Reading

- European Regulations and Guidance
  Consolidated Veterinary Medicines Directive 2001/82/EC, Title III
  Regulation (EC) No. 726/2004, Title III, Chapter 1
  Maximum residue limits - Regulation (EEC) No. 2377/90 & Council Regulation (EC) No. 1308/1999
  The Rules Governing Medicinal Products in the European Union, Volume 6 – Notice to Applicants, Veterinary Medicinal Products
  The Rules Governing Medicinal Products in the European Union, Volume 8 – Maximum Residue Limits
  http://ec.europa.eu/enterprise/pharmaceuticals/eudralex/index.htm.
- US Code of Federal Regulations
  Veterinary biologics – 9 CFR Parts 102,103
  www.access.gpo.gov.
  Veterinary drugs – 21 CFR Parts 511 & 514
  www.fda.gov.
- VICH Guidelines
  www.vichsec.org.

8
Variations to the Drug Authorisation Process

8.1
Chapter Introduction

The previous chapters have examined the process for development and authorisation of a drug product containing a new active ingredient. This represents the most tortuous path to market for any medicinal product. This chapter looks at the variations to the authorisation process that can apply depending on the nature of the product or its intended use. These include incentives for orphan drugs and paediatric applications, alterations or extensions to existing approved drug products, and the authorisation of generic copies of approved drugs and other special authorisation procedures.

8.2
Provisions in Support of Special Drug Applications

The regulations contain certain incentives and provisions to encourage and support the development of drugs for treatment of certain categories of disease or condition. Principal among these are Orphan Drug regulations, which are intended to encourage the development of drugs for rare diseases and various provisions that facilitate accelerated access to significant new therapies.

8.2.1
Orphan Drugs

Products intended to treat rare diseases or conditions which, considering the size of the target market are not commercially attractive, may be categorised as Orphan Drugs. Incentives to encourage the development of such products were introduced in the USA in 1983 with the enactment of the Orphan Drug Act. It was not until Regulation (EC) No. 141/2000 and Regulation (EC) No. 847/2000 were adopted that the EU provided similar incentives.

Medical Product Regulatory Affairs. John J. Tobin and Gary Walsh
Copyright © 2008 WILEY-VCH Verlag GmbH & Co. KGaA, Weinheim
ISBN: 978-3-527-31877-3

To avail themselves of the incentives, the sponsor must first of all request that the drug is designated as an orphan product by the relevant authorities, the Committee for Orphan Medicinal Products (COMP) in the EU and the Office of Orphan Products Development (OOPD) in the US. The criteria on which orphan status is granted differ slightly between the EU and the US. In Europe, consideration is given to the nature and prevalence of the disease, the commercial attractiveness and the availability of existing therapies, whereas US designations are just based on incidence and commercial viability. The specific designation criteria are shown in Figure 8.1. The US authorities will accept a specific paediatric application as a plausible subset for orphan status designation, which may be granted separate from the overall drug status.

Market exclusivity has proved to be the most significant benefit that can be gained from obtaining orphan status. This is intended to provide the sponsor with a marketing window to recover the costs of the development. In Europe, the authorities will not grant authorisation to similar drugs for the same therapeutic indication for a period of 10 years after approval of the first drug, unless the new drug can be shown to be clinically superior. Similar drugs include active ingredients with the same principal molecular structures and mechanism of action, but with possible variation

EU Orphan Drug Criteria
A medicinal product shall be designated as an orphan medicinal product if its sponsor can establish:

(a) that it is intended for the diagnosis, prevention or treatment of a life-threatening or chronically debilitating condition affecting not more than five in 10 thousand persons in the Community when the application is made,
or
that it is intended for the diagnosis, prevention or treatment of a life-threatening, seriously debilitating or serious and chronic condition in the Community and that without incentives it is unlikely that the marketing of the medicinal product in the Community would generate sufficient return to justify the necessary investment;

and

(b) that there exists no satisfactory method of diagnosis, prevention or treatment of the condition in question that has been authorised in the Community or, if such method exists, that the medicinal product will be of significant benefit to those affected by that condition.

US Orphan Drug Criteria (Rare Disease)
"rare disease or condition" means any disease or condition which

(a) affects less than 200 000 persons in the United States,
or
(b) affects more than 200 000 in the United States and for which there is no reasonable expectation that the cost of developing and making available in the United States a drug for such disease or condition will be recovered from sales in the United States of such drug.

Figure 8.1 Criteria for designation as an Orphan Drug.

due to isomerism, salts, complexes, conjugates or minor structural alteration. The FDA will not approve other versions of the product for orphan-designated use for a period of 7 years after the first approval, provided that an adequate supply of the product is maintained on the US market.

Financial and other benefits are also available during the development of the drug. Sponsors in the US may receive tax credits of up to 50% against expenses incurred in clinical trials. Grant aid is available in both the US and EU to support research into the development of Orphan Drugs. The regulatory agencies will also provide assistance in developing clinical trial protocols, since well-designed trials are essential in a situation where there may be a limited availability of trial subjects. The OOPD acts as a coordinating ombudsman to facilitate efficient review by the relevant centre. In Europe, all applications are reviewed by the COMP. Finally, sponsors in the US are exempt from the Prescription Drug User Fees that are normally paid to the FDA for regulatory reviews. In Europe, the European Medicines Agency (EMEA) will grant a rebate on the application fee, depending on the availability of funds in the budget specifically allocated by the Commission for this purpose.

8.3
Accelerated Access to New Drug Therapies

The time required to fully establish the safety and efficacy of a new drug in the eyes of the regulatory authority might not always be to the benefit of the proposed patient group. This is particularly so in the case of drugs intended to treat serious or life-threatening diseases where no effective therapy currently exists. Various provisions have been incorporated into the regulations to allow accelerated access to such therapies.

8.3.1
EMEA Accelerated Review

In Europe, a sponsor may request an accelerated review of a marketing authorisation by the EMEA on grounds that the product is of major interest to public health, particularly from the viewpoint of therapeutic innovation. If granted, an accelerated review must be conducted within 150 days as opposed to the 210 days allowed for a standard technical assessment.

8.3.2
Compassionate Use

Member States may also permit access to non-approved drugs on compassionate grounds so that patient groups who suffer from chronic, seriously debilitating or life-threatening diseases for which no satisfactory authorised drugs exist, may receive treatment while the drug is still undergoing clinical trials or awaiting marketing authorisation. The Member State must inform the EMEA if such use is contemplated. After consideration, the CHMP may issue recommendations on the conditions of use, distribution, or the patient groups targeted.

8.3.3
Fast-Track Products

Drugs that are intended to treat serious or life-threatening diseases and that demonstrate potential meaningful therapeutic benefit over existing treatments may be designated as *fast-track products* under US regulations. The FDA will then take appropriate measures to facilitate the development and accelerated review of such products. The FDA may grant an authorisation on the basis of clinical trials that demonstrate effects on surrogate end points or clinical indicators, without having to wait for confirmation of the final therapeutic effect (e.g. prevention of morbidity). Authorisations are usually conditional on the sponsor continuing the trials to fully validate the therapy.

8.3.4
Treatment INDs

A Treatment Investigational New Drug Application (IND) is the US mechanism that enables patients suffering from serious or immediately life-threatening diseases to gain access to non-approved drugs, when no satisfactory authorised therapy exists. Drug administration must follow a treatment protocol submitted to the FDA as part of the IND application. Treatment INDs are usually only granted while the drug is undergoing Phase III clinical trials or is at the NDA evaluation and authorisation stage.

8.3.5
Paediatric Applications

US regulations require that, by default, the sponsor must specifically assess the safety and effectiveness of a drug in paediatric patients. However, waivers will be granted if it can be justified that the drug does not have meaningful therapeutic application in children, or is unlikely to have benefit over existing therapies, or where it is not possible to undertake studies because of the numbers of available patients. A paediatric application may be submitted separately, subject to FDA agreement. As a reward, drugs with specific paediatric indications for use are granted 6 months extra market exclusivity over the standard terms that would otherwise apply.

Recently, similar legislation has been introduced in the European Union under regulations EC/1901/2006 and EC1902/2006. This requires that marketing authorisation applications for new drugs submitted after the 28 July 2008 must be accompanied by either the results of specific studies demonstrating safety and effectiveness in the paediatric populations or waivers or deferrals granted by the European Medicines Agency. A new Paediatric Committee has been established to provide advice on paediatric issues, to adopt opinions on paediatric investigation plans, which must now be submitted to the agency before commencing the studies, and to grant deferrals or waivers where appropriate. The committee will also assist in reviewing the paediatric section of a marketing authorisation application. As reward for undertaking an agreed paediatric investigation plan, manufacturers may gain additional market protection ranging from six months to two years in the case of an orphan medical product. It is also possible for manufacturers of existing drugs that no longer enjoy any protection to submit new 'paediatric use' marketing authorisation applications and thus obtain the data/market exclusivity periods associated with a standard authorisation. Provision is also made for financial aid to conduct the research and development of paediatric drugs.

8.4
Approval of New Drugs when Human Efficacy Studies are not Ethical or Feasible

US regulations recognise that trials with human subjects may not be ethical for drugs intended to prevent or ameliorate serious or life-threatening conditions caused by exposure to lethal or permanently disabling toxic, biological, chemical, radiological or nuclear substances. In such circumstances, the FDA may approve applications based on well-designed safety and efficacy studies in animals only.

8.5
Animal Drugs for Minor Use and Minor Species

The Minor Use and Minor Species (MUMS) amendments to the FD&C Act (Sections 571, 572, 573) were signed into US law in 2004. Their purpose is to make drugs legally available for use in minor species, or for the treatment of rare diseases in major species. Major animal species are defined as cattle, horses, swine, chickens, turkeys, dogs and cats. All other species are considered minor species. A drug developer can benefit from MUMS provisions in one of three ways.

8.5.1
Conditional Approval

The developer may receive *conditional approval* for up to five years, through annual renewals, on the basis that the safety has been established, but that complete efficacy data will only be generated during the conditional approval period. The labelling must bear the statement "conditionally approved by FDA pending a full demonstration of effectiveness under application number".

8.5.2
Indexing

In certain cases, such as the treatment of zoo animals or endangered species, it may be impossible to recover the cost of approval of a drug for that species. In such circumstances, the FDA can add the drug to an *index of legally marketed non-approved drugs*. A drug can only be indexed for use in situations where there is no risk of it entering the human food chain.

8.5.3
Designation

A developer may apply to have a new drug designated. *Designation* entitles the developers to avail themselves of the same type of incentives in terms of supports and market exclusivity that apply to human orphan products.

8.6
Use of Non-Authorised Drugs for Animal Treatment in the EU

The EU regulations permit veterinarians to treat animals with animal drugs that have not been authorised for that species or, alternatively to use human drugs, in situations where there is no authorised drug for treatment of the condition in the species.

8.7
Changes to an Authorised Drug

The regulatory authorities must be informed of all changes relating to an authorised drug. The procedure for processing changes will vary depending on the magnitude of the change. Prior authorisation is generally required for changes designed to alter the drug product or to extend its therapeutic use (i.e. changes to the active ingredients, strength, form, route of administration, dosage regimen or therapeutic indication), or manufacturing changes that have the potential to significantly alter the safety or efficacy of the product. Notification is usually sufficient for other minor changes that are unlikely to have an adverse effect on the safety or efficacy of the product. Different approaches to the categorisation and procedures for processing changes apply in Europe and the US.

8.8
EU System for Processing Changes

Changes to an authorised drug in the EU may be categorised as *Extensions, Major Variations* (Type II variations) or *Minor Variations* (Type IA or IB). All variations or extensions relating to the basic active ingredient are considered part of the same "global" marketing authorisation. Commission Regulations (EC) No. 1084/2003 and

(EC) No. 1085/2003 outline the procedures for processing variations to marketing authorisations granted via National and Community procedures, respectively. The Annexes to the regulations provide criteria for categorisation of changes.

8.8.1
Extension Applications

Extension Applications are applicable to the following changes, which are further detailed in Annex II to the regulations:

- Changes to the active substance(s), such as use of different salts, complexes, isomers, or slightly different biological materials.
- Changes to strength, pharmaceutical form or route of administration.
- Other changes specific to veterinary medicinal products to be administered to food-producing animals, such as changes in target species.

One exception to the above criteria is the annual renewal procedure for human influenza vaccines, which can be processed using a Type II Variation Application, even though it involves a change in the active substance – that is, the strains of virus used to prepare the vaccine. Extension Applications are processed in accordance with the standard procedures for granting a marketing authorisation. Applications must be submitted to the EMEA (Community authorised drugs) or Member State Competent Authorities (National authorisations) as appropriate, together with a dossier containing new information supporting the extension. The applicant can refer to the original dossier for other information relevant to the extension.

8.8.2
Major Variation (Type II)

A Major or Type II Variation applies to any change that is not categorised as either an Extension or a Minor Variation. Changes to introduce a new therapeutic use are processed as Major Variations. Applications must be submitted to the EMEA in the case of drugs that hold a Community authorisation or to the all the Competent Authorities of the Member State where the drug is authorised in the case of drugs that were authorised by national Competent Authorities. Applications must be accompanied by relevant documents and data in support of the variation. Variations in the indications for use shall be supported by clinical and pre-clinical data, if justified. The relevant EMEA committee or the reference Member State as appropriate must deliver an opinion within 60 days of verification that the application has been submitted correctly. This period may be extended to 90 days in the case of new indications for use or the addition of new animal species, but this still represents a substantially shorter review period than is allowed for a standard marketing autho-risation or extension application. Concerned Member States have 30 days to recog-nise the opinion of the reference Member State, while the Commission must update a Community decision in accordance to centralised procedures. Shorter review periods (30 or 45 days) apply to flu vaccines that follow this procedure, while urgent safety restrictions may be processed within 24 h.

8.8.3
Minor Variation (Type IA or IB)

A list of minor variations is detailed in Annex I to the regulations. This contains 46 types of minor variations that include changes in the name and address of a marketing authorisation holder, manufacturing sites, batch size, test procedures, specifications, excipients, shelf life, packaging or labelling. The individual variations are categorised as either Type IA, which are of lowest significance, or as Type IB, which are slightly more significant. The relevant authorities must be notified of such variations, but specific authorisation is not required. For a Type IA variation, the authorities will just check that all the submitted documents meet the requirements. They must inform the marketing authorisation holder of the validity of the notification within 14 days. The authorities are allowed 30 days to consider a valid Type 1B notification, during which period they must inform the holder of any objections. Otherwise, the variation may be introduced once the 30 days have elapsed.

8.9
Processing Changes in the US

The procedure for introducing changes to an approved drug in the US is dictated by type of data that must be submitted to support the alteration. Changes that require review of new investigations to approve the modification, other than bioavailability or bioequivalence studies, must be submitted as supplementary NDAs/NADAs, as they are viewed as new drugs, meaning that are not generally recognised (by the FDA) as being safe and effective for the indications proposed. These include changes to the active ingredient, pharmaceutical form, strength, or route of administration of the drug product, or alterations to the dosage regimen, or addition of new therapeutic indications for use. The requirement to submit a new therapeutic use using new drug approval procedures differs from the EU system, where it can be processed as a Major variation. The applicant is only required to submit new data that support the modification, and can refer to the original submission for other relevant data, provided that they are the owner or have obtained a right of reference.

8.9.1
Manufacturing Change Supplements

Changes to manufacturing processes, specifications, manufacturing sites, or packaging, that have the potential to have an adverse effect on the identity, strength, quality, purity or potency of the drug product or labelling, or other miscellaneous changes that might otherwise influence the safety or effectiveness of the drug, must be notified as *manufacturing change supplements*. While such changes are not intended to significantly alter the drug or to extend its use, the magnitude of potential effects dictate how these supplements are categorised and processed. The FDA has issued specific guidance for industry to assist in selecting the appropriate supplement category.

In each case, the approved drug holder must submit sufficient evidence and information to enable the FDA understand and assess the impact of the proposed change.

8.9.2
Major Changes

Major changes require review and prior approval by the FDA before they can be implemented. Examples of changes that would be considered major include the following:

- Quantitative or qualitative changes to the drug product
- Changes that may affect product sterility
- Changes to the synthesis or manufacture process that may affect the impurity profile
- Changes that require bioequivalence validation
- Changes to labelling resulting from post-approval studies.

8.9.3
Moderate Changes

There are two categories of moderate change that can be submitted to the FDA, namely "changes being effected in 30 days" and "changes being effected". The applicant must wait for 30 days before implementing a 30-day change, whereas a change being effected can be implemented once the FDA have been notified. Changes that would be considered suitable for a 30-day notification supplement include the following:

- Changes to a closure system that does not affect quality of the drug product
- Changes in final process scale involving new equipment
- Relaxation of acceptance criteria or deletion of tests in accordance with official compendia.

Immediate changes may include the following:

- Changes to specifications or test methods intended to provide increased assurance as regards product quality
- Changes to the size or shape of a container for a non-sterile drug
- Changes to labelling to strengthen statements on contraindications, precautions, overdose, administration, etc.

8.9.4
Minor Changes

Minor changes do not have to be notified to the FDA by a specific supplement, but may be included in the next Annual Report. Minor changes would include editorial corrections to labelling, deletion or reduction in an ingredient intended to affect the colour, replacement of equipment with similar equipment, or the extension of shelf life based on real-time data.

8.10
Authorisation of Generic Drugs

Once the patent protection on an authorised new drug has expired, other manufacturers will wish to introduce generic copies of the drug. The new manufacturer must apply for a marketing authorisation for the generic drug. However, the applicant will not have to submit pre-clinical and clinical data, if bioequivalence with the approved reference drug can be demonstrated (demonstrating bioequivalence entails showing that two products are pharmaceutically equivalent to each other). The authorities will authorise the generic copies on the basis that the safety and efficacy established for the reference drug will also apply to the copy when both versions of the drug display similar bioavailability profiles. While this strategy may appear to offer an unfair advantage to the generic manufacturer, there are a number of reasons why it is in the interest of public health. The authorities do not have to waste time reviewing replicate pre-clinical and clinical data, and instead are able to allocate more of their resources to the review of new drugs, which benefits the innovator who will be able to get to market more quickly. Unnecessary testing on human and animal subjects is avoided. Likewise, the cost of bringing a generic drug to market is reduced, thus enabling prices to be reduced, which in turn makes the product more accessible to the public. The specific criteria for accepting generic copy submissions differ slightly between Europe and the US.

8.10.1
EU Regulations

The authorisation of generic medicinal products is covered by Article 10 of the Human Medicines Directive 2001/83/EC and Article 13 of the Veterinary Medicines Directive 2001/82/EC. A generic drug is defined as a medicinal product that has:

- the same qualitative and quantitative composition in active substances as the reference product;
- the same pharmaceutical form as the reference medicinal product; and
- and whose bioequivalence with the reference medicinal product has been demonstrated by appropriate bioavailability studies.

The different salts, esters, ethers, isomers, mixtures of isomers, complexes or derivatives of an active substance shall be considered to be the same active substance, unless they differ significantly in properties with regard to safety and/or efficacy, in which case additional safety and efficacy data are required. The "same qualitative and quantitative composition" only applies to the active ingredients. Differences in excipients will be accepted unless there is concern that they may substantially alter the safety or efficacy. The "same pharmaceutical form" must take into account both the form in which it is presented and the form in which it is administered. Various immediate-release oral forms, which would include tablets, capsules, oral solutions and suspensions, shall be considered the same pharmaceutical form for this purpose.

Although not falling within the definition of a generic drug, an applicant may introduce new therapeutic uses, strengths, or pharmaceutical forms as part of a hybrid submission, provided that the changes from the reference product are supported by appropriate pre-clinical and clinical data, and residue testing in the case of veterinary products. Similarly, pre-clinical and clinical data may be submitted to bridge the safety and efficacy gaps in the case of similar biological products, which may not be identical due to different raw materials or manufacturing processes, or generic drugs where bioequivalence could not be established. Applications may be submitted to the EMEA if the reference product was centrally approved. Otherwise, applications must be submitted to Member States for authorisation via either the Decentralised or Mutual Recognition procedures. A member state may authorise a generic drug even if the reference drug was never authorised in that state.

8.10.2
US Regulations

Requests for the approval of generic drugs in the US may be submitted as an Abbreviated New Drug Application (ANDA) or Abbreviated New Animal Drug Application (ANADA). Such submissions rely on the approved application of another drug with the same active ingredient to establish safety and efficacy. The criteria under which an abbreviated submission can be made are more restrictive than the equivalent regulations in Europe. The generic drug should normally be identical to an approved drug in terms of active ingredient(s), dosage form, strength, route of administration, and conditions of use. However, the applicant may petition the FDA for permission to submit an abbreviated application for a drug product which is not identical to a listed drug in terms of route of administration, dosage form, and strength, or in which a single active ingredient in a listed combination drug is substituted by another similarly acting ingredient. New indications for use or different salts, esters, ethers, isomers, mixtures of isomers, complexes or derivatives of an active substance or other modifications supported by clinical safety and efficacy studies may only be submitted as NDAs or NADAs.

8.11
Reference Drug Exclusivity

In order to encourage and reward the originators of new drugs the regulations contain restrictions to prevent the entry of generic copies onto the market for set time periods.

In Europe, a manufacturer may not submit an application that relies on a reference drug (unless granted a right of reference by the originator) for a period of 8 years after the initial approval of the reference drug – this is the *data exclusivity period*. Such a product may not be placed on the market for a further 2 years, giving 10 years total market exclusivity. Market exclusivity for a human medicine may be extended to 11 years if a new therapeutic indication with significant benefit over existing therapies was approved within the 8-year data exclusivity window. The market

Table 8.1 Protection intervals (years) allowed to developers of innovator products.

Type of protection	EU	US
Patent	20	20
Maximum patent extension	15 from first marketing approval	14 from marketing approval
Data exclusivity	8	5
Market exclusivity	10	–
Orphan drug	10	7
New therapeutic indication	+1 (human) to +3 (veterinary)	3 from approval of indication
Paediatric applications	+0.5	+0.5

exclusivity of veterinary medicines may be extended to a maximum of 13 years by adding new species, or where the product is intended for fish or bees.

In the US, a manufacturer may not submit an application that relies on data from an approved submission for a period of 5 years after the reference product was authorised, unless the "owner" of the original data has granted a right of reference. The FDA may not approve an application that relies on data submitted as a supplement (e.g. new therapeutic indication) to an authorised drug for a period of 3 years after approval of the supplement. These exclusivity periods will be extended by 6 months if the reference submissions contain paediatric data. The FDA will also not

Development Scenario

- New pharmaceutical form approved 6 years after approval of the original product.
- A paediatric application approved for the new pharmaceutical form 7 years after approval of the original product.
- New dosage regimen for the original product approved 8 years after approval of the original product.

EU Relevant Exclusivity

- The generic manufacturer may not place a copy of either the original product or the new pharmaceutical form on the market for a period of 11 years after the approval of the original product (i.e. 10 years + 1 year added to the global market exclusivity period, because of new indications).

US Relevant Exclusivity

- The generic manufacturer may not submit an ANDA for the original product until 5 years after approval of the original form.
- The generic manufacturer may not submit an ANDA for the new pharmaceutical form until 11.5 years after approval of the original product (i.e. 5 years from the date of approval of the new form + 0.5 years for the paediatric application).
- The new dosage regimen may not be approved until 11 years after the original product was approved (i.e. 8 + 3).

Figure 8.2 Illustration of the application of exclusivity periods according to EU and US regulations.

grant approval to a generic product if the drug is still under patent protection. This is why US applications must include information on patents and, in the case of abbreviated applications, certification as to patent status – that is, expiry date or claims of non-infringement or invalidity.

The protection periods available to originators are summarised in Table 8.1 and Figure 8.2.

8.12
Other Authorisation Procedures

There are a number of other scenarios where medicinal products may be brought to market exempt from the requirement to submit applications supported by full pre-clinical and clinical data.

8.12.1
Well-Established Medical Use Products

The EU will accept applications without supporting pre-clinical and clinical data, if it can be demonstrated that the active substances have been in well-established medical use in the Community for at least 10 years, with recognised efficacy and an acceptable level of safety. This route would be appropriate for many common over-the-counter (OTC) products. Safety and efficacy is supported by providing copies of published scientific literature as part of the submission; that is, the submission relies on safety and efficacy data available in the public domain, as opposed to confidential data from authorised applications that is the cornerstone of generic applications.

8.12.2
Combination Products

An EU application for a combination product that contains previously authorised active substances must be supported by appropriate safety and efficacy data for the combination. However, there is no need to submit data on the safety and efficacy of the components individually. The combination product is considered a unique product distinct from the authorisations granted to the individual ingredients, and will thus have a separate exclusivity clock running from the date when the combination was authorised.

8.12.3
Homeopathic Medicines

Homeopathic medicines may avail of a simplified registration procedure if they fulfil the following criteria:

- They are administered orally or externally (via European Pharmacopoeial route in the case of veterinary products).

- No specific therapeutic indication appears on the labelling.
- There is a sufficient degree of dilution to guarantee the safety of the medicinal product (not more than one in 10 000 of the mother tincture or 1/100th of the smallest dose requiring a medical prescription for an allopathic medicinal product).

A dossier that focuses on quality (manufacture, control, stability) and labelling of the product must be submitted in support of the application.

8.12.4
Traditional Herbal Medicines

Traditional herbal medicines may also avail of a simplified registration procedure if they fulfil the following criteria:

- They have indications exclusively appropriate to traditional herbal medicinal products which, by virtue of their composition and purpose, are intended and designed for use without the supervision of a medical practitioner for diagnostic purposes or for prescription or monitoring of treatment.
- They are exclusively for administration in accordance with a specified strength and posology.
- They are an oral, external and/or inhalation preparation.
- They must have been in traditional use for at least 30 years, 15 years of which must be in the EU.
- Traditional use data are available that supports the precepts that they are not harmful under the conditions of use, and that efficacy is plausible based on longstanding use and experience.

The approach is quite similar to that for a well-established medical product, in that the applicant must submit bibliographic data supporting the safety, efficacy and duration of the traditional use of the herbal remedy (see Table 8.2).

All other homeopathic and herbal medicines that do not qualify for simplified registration procedures will have to go through the standard authorisation procedure with supporting technical dossiers. The Committee for Herbal Medicinal Products has been charged with preparing monographs on herbal products so as to further facilitate the registration of traditional and well-established use herbal products.

8.12.5
US Regulation of OTC Drugs

OTC drugs are not required to go through the new drug application procedures that apply to prescription-only ("Rx") drugs in the US. Instead, they can be marketed on the principle that all the active ingredients have been deemed by the FDA to be "Generally Recognised as Safe and Effective" (GRASE). To be available without prescription, OTC drugs must have a wide safety margin, which is usually underpinned by a long history of safe use. As many of the drugs came on the market before there was a requirement to demonstrate safety and efficacy, the FDA undertook a

Table 8.2 Overview of study requirements and data presentation for different types of application.

Application type	Labelling	Quality	Safety	Efficacy
Standard authorisation	Yes (SPC)	Yes (CMC and Stability)	Yes (Pre-clinical and Clinical data)	Yes (Clinical data)
Generic authorisation	Yes (SPC)	Yes (CMC and stability)	No (Reference dossier)	No (Reference dossier)
Well established use authorisation	Yes (SPC)	Yes (CMC and stability)	No (Published literature)	No (Published literature)
Homeopathic registration	Yes	Yes (Manufacture, control and stability)	No (Sufficient dilution)	No (No therapeutic indications)
Traditional Herbal Product Registration	Yes (SPC)	Yes (Manufacture, control and stability)	No (Published literature)	No (Published literature)

review of OTC products in 1972. Rather than look at what was then in excess of 300 000 individual products, expert panels reviewed them by active ingredient and therapeutic class. These reviews were based on the publicly available scientific and medical literature as to safety and efficacy. Following public consultation, these reviews were condensed into monographs, which define the basic formulation, use and labelling conditions that a manufacturer must follow. Manufacturers may submit an amendment to a monograph to switch a more recently developed drug to OTC status.

8.13
Chapter Review

Depending on the circumstances, manufacturers may be able to avail themselves of variations from the standard marketing authorisation procedures. These include incentives such as the Orphan Drug, paediatric use and minor species minor use provisions, which are intended to encourage greater availability of drugs for treatment groups that might not be financially attractive. In order to speed up access to drugs that represent a significant advance in the treatment of serious/life-threatening diseases, there is the possibility to fast track the approval and/or make them available on compassionate grounds prior to approval. Manufacturers that wish to introduce generic copies of innovator products, are not required to repeat non-clinical and clinical studies, but can rely on the original innovator's data. The innovator may use a combination of patent and data protection to exploit the drug before generic copies can come on the market. The regulatory authorities must be informed of all changes relating to an approved product. This can range from the most minor change, which does not require prior approval, to significant changes that must follow essentially the same approval process as the original product. The specific requirements differ somewhat between Europe and the US.

8.14
Further Reading

- European Regulations and Guidance
  Consolidated Veterinary Medicines Directive 2001/82/EC, Title III
  Consolidated Human Medicines Directive 2001/83/EC, Title III
  Orphan Drug Regulations (EC) No. 141/2000 and (EC) No. 847/2000
  Paediatric Use Regulations (EC) No. 1901/2006 and (EC) No. 1902/2006
  Variations Regulations (EC) No. 1084/2003 and (EC) No. 1085/2003
  The Rules Governing Medicinal Products in the European Union, Volume 2, Notice to Applicants, Medicinal Products for Human Use
  The Rules Governing Medicinal Products in the European Union, Volume 6, Notice to Applicants, Veterinary Medicinal Products
  http://ec.europa.eu/enterprise/pharmaceuticals/index_en.htm

- US Code of Federal Regulations
  21 CFR Part 314
  21 CFR Part 514
  21 CFR Part 601
- US Food, Drug & Cosmetic Act
  Chapter 5, Sub chapter B, F

9
Medical Devices

9.1
Chapter Introduction

Medical devices cover the most diverse array of products in the healthcare manufacturing sector. They range from simple devices such as bandages, tooth-brushes or spectacles, through life-maintaining active implantable devices such as heart pacemakers, to the most sophisticated diagnostic imaging and minimally invasive surgical equipment. There is also considerable variation in the underlying technology applied. Devices can rely on any combination of mechanical, electronic or chemical/biochemical action to achieve their purpose. Thus, in contrast to drugs, a more flexible regulatory approach is required in order to adequately address safety and effectiveness concerns for this product sector. Regulatory strategies are built around the classification of devices according to risk, a corresponding tiered application of regulatory requirements, and the use of appropriate standards to demonstrate compliance. This chapter examines the application of these principles to the development of devices.

9.2
Regulatory Strategy for Medical Devices in the EU

Harmonised regulation of devices was introduced in the EU during the 1990s. Most medical devices are regulated via the Medical Devices (MDs) Directive 93/42/EEC. This covers all medical devices with the exception of Active Implantable Medical Devices (AIMDs) and In Vitro Diagnostics (IVDs), which are regulated separately via directives 90/385/EEC and 98/79/EC. In order to achieve efficient harmonisation across Member States, the directives employ a common strategy of just specifying essential requirements that devices must meet. As such, the directives only define legal requirements as regards the end results or goals that must be attained, but allow flexibility as to the methods or solutions that a manufacturer may employ in achieving these goals. The fundamental objective behind the essential requirements is to ensure that only those devices that are safe and effective for their intended use may be

Medical Product Regulatory Affairs. John J. Tobin and Gary Walsh
Copyright © 2008 WILEY-VCH Verlag GmbH & Co. KGaA, Weinheim
ISBN: 978-3-527-31877-3

marketed. Details of some of the essential requirements differ between the three directives, but the key elements can be summarised as shown in Figure 9.1.

When the directives were introduced, they were applied to both new and existing devices. Thus, all device manufacturers are required to establish conformity with essential requirements, so that they can apply the CE Mark of conformity and legally market their products in the EU.

The CE Mark is not specific to medical devices, but is used generally to indicate to European consumers that a product conforms to applicable European performance and safety requirements. It can be found on electrical equipment, children's toys and safety equipment, among other products.

The basic mark is illustrated in Figure 9.2. Depending on the type of product, it may be accompanied by the identification number of the Notified Body responsible for performing specified conformity assessment tasks (see Chapter 10).

The strategy of reliance on achieving end results as opposed to specifying the methods for achieving the results was vital to ensuring a smooth transition to compliance with the directives, particularly considering the diversity in the types of devices and technologies that exist. So also was the fact that the directives allow for flexibility in the conformity assessment procedures that may be used to establish compliance with the requirements. This was particularly important for older devices, which may have been developed before accepted design practices or norms were available.

9.2.1
Use of Standards to Establish Conformity

Although the directives allow flexibility in the methods and solutions that a manufacturer may employ to fulfil the essential requirements, they promote the use of standards as the preferred method to establish conformity. Specifically, the directives stipulate that Member States shall presume compliance with the essential requirements in respect of devices that are in conformity with relevant harmonised standards.

At the European level, standards are developed through the European Committee for Standardization (CEN), or the European Committee for Electrotechnical Standardization (CENELEC) in the case of electrical standards. These are comprised of representatives from the national standards organisations of each Member State, plus Switzerland, Norway and Iceland. Draft standards are prepared, usually with input from relevant industry associations, and circulated to the member bodies for comment. Following the inclusion/rejection of comments, final drafts are adopted as European Norms (EN) by majority vote, using a weighted voting system. Member bodies must then adopt or reference the standards as national standards without alteration and abandon any pre-existing national standards that may conflict with the international standard.

Standards may be developed at global level through the International Standards Organisation (ISO) or the equivalent International Electrotechnical Commission (IEC) for electrical equipment. These contain representatives from standards bodies

Essential Requirements

I General Requirements

1. Safety
The devices must be designed and manufactured in such a way that when used as intended they do not compromise the safety of patients or users. Any risks that may be associated with their use must be acceptable when weighed against the benefits to the patient and be compatible with a high level of protection of health and safety.

2. Effectiveness
The devices must be designed and manufactured in such a way that they will deliver their intended performance under the recommended condition for use.

3. Risk Management
The design and construction of devices must adhere to safety principles by applying the following measures in the following order:
— eliminate or reduce risks as far as possible (inherently safe design and construction),
— where appropriate take adequate protection measures in relation to risks that cannot be eliminated,
— inform users of the residual risks due to any shortcomings of the protection measures adopted.

4. Stability
The devices must retain their performance characteristics over their claimed lifetime under the normal stresses associated with their use.
The devices must be designed manufactured and packaged so as to maintain their performance characteristics under the intended conditions of transport and storage.

II Design and Construction/Manufacturing Requirements
Consideration should be given as appropriate to the following aspects.

6. Choice of Materials
Toxicity or flammability of materials
Compatibility with biological tissues or other materials
Measures to prevent infection from devices containing materials of biological origin

7. Devices utilising a power source
Protection against electrical shocks (device and or connections)
Interference with or by other devices (electromagnetic interference, etc.)
Reliability of power source (AIMDs) and or software
Protection against thermal or mechanical risks

8. Devices with a measuring function
Accuracy and stability of results
Use of SI units
Traceability to reference materials/standards

9. Protection against radiation (lasers, X-ray, etc.)
Measures to limit exposure of users or patients to radiation.
Adequate control features and instructions to ensure safe administration of intended exposure.

10. Sterile devices
Packaging and validated sterilisation procedures to ensure sterility

11. Devices used in combination with other devices or equipment
Performance of the overall system must be considered

12. Information supplied by the manufacturer
Mandatory information on labels and instructions for use (specified according to directive/type of device)
Use of symbols on labels

Figure 9.1 Summary of key elements of the Essential Requirements from the medical device directives 90/385/EEC, 93/42/EEC and 98/79/EC.

C E **Figure 9.2** The CE Mark.

worldwide. The process for developing standards is similar to that at European level. A 75% vote in favour is required before a standard may be adopted as an International Standard (IS). Where no appropriate international standards exist, manufacturers may rely on relevant national standards. For example, standards established by the US National Committee for Clinical Laboratory Standards (NCCLS) are widely used by manufacturers worldwide for establishing the performance of IVD reagents.

The European Commission has mandated CEN/CENELEC to develop many standards to support manufacturers in the realisation of the essential requirements of the Directives. If the Commission determines that an adopted standard is fit for this purpose, then a reference is published in the Official Journal of the EU, which elevates the status of the standard to that of a harmonised standard for European regulatory purposes.

The Commission provides a list of over 230 standards that may be used to establish conformance with the essential requirements of the medical device directives.

The text of the standards is not published in the Official Journal, as they are protected by copyright and must be purchased through sales outlets of the standardisation bodies.

The single most important standard that has relevance for all medical device manufacturers is ISO 13485:2003, Medical devices – Quality management systems – Requirements for regulatory purposes. It is based on the more familiar ISO 9001:2000 standard, Quality management systems – Model for quality assurance in design, development, production, installation and servicing, that is used by organisations in general. However, it differs from ISO 9001 in that its primary objective is to provide a "harmonized" model for quality management system requirements that satisfy international medical device regulations. Consequently, its main focus is on maintaining the effectiveness of the quality management system as opposed to ISO 9001, where continuous improvement and customer satisfaction are fundamental drivers of the system. This stems from the precept that, in accordance with regulatory requirements, medical devices must be safe and effective for their intended purpose before being placed on the market, and thus product improvement or customer satisfaction should be less of an issue after the launch. There is also a strong emphasis on risk management to ensure that the devices are safe and effective. In common with ISO 9001, ISO13485 applies a Plan-Do-Check-Act (PDCA) methodology to processes, as illustrated in Figure 9.3. The main headings of the standard are outlined in Figure 9.4.

9.2.2
Classification of Devices

The range of devices regulated under the general medical device directive, 93/42/EEC can vary from the simple devices that pose little or no risk to patient or user health to those that are life-critical. It is neither economically feasible, nor justifiable in practice

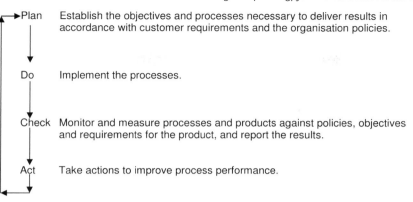

Plan   Establish the objectives and processes necessary to deliver results in accordance with customer requirements and the organisation policies.

Do   Implement the processes.

Check   Monitor and measure processes and products against policies, objectives and requirements for the product, and report the results.

Act   Take actions to improve process performance.

Figure 9.3 The PDCA model.

4. Quality Management System
4.1 General Requirements (Implement quality system and continually improve it)
4.2 Documentation Requirements (Quality Manual, control of documents and records)

5. Management Responsibility
5.1 Management Commitment (Show evidence of top management commitment)
5.2 Customer Focus (Meet the requirements and satisfy your customers)
5.3 Quality Policy (Have top management express quality intentions)
5.4 Planning (Set quality objectives and plans to meet objectives)
5.5 Responsibility, Authority, and Communication (Define duties, appoint a Quality Manager, communicate effectiveness of system)
5.6 Management Review (Review performance and effectiveness of system)

6. Resource Management
6.1 Provision of Resources (Provide necessary resources to meet requirements)
6.2 Human Resources (Competence and training)
6.3 Infrastructure (Provide facilities, equipment, and support services)
6.4 Work Environment (Manage combination of human and physical factors)

7. Product Realization
7.1 Planning of Product Realization (Plan and develop processes to produce product)
7.2 Customer-Related Processes (customer requirements, orders and communication)
7.3 Design and Development (Design Plans, Inputs, Outputs, Reviews, Verification, Validation and Change)
7.4 Purchasing (Evaluate suppliers, verify supplies)
7.5 Production and Service Provision (Control of production, process validation, product status)
7.6 Control of Monitoring and Measuring Devices (Calibrate measuring equipment for valid results)

8. Measurement, Analysis and Improvement
8.1 General (Plan, measure, analyse, and improve processes)
8.2 Monitoring and Measurement (customer feedback, internal audits, process monitoring, finished product release)
8.3 Control of Non-conforming Product (Prevent use or delivery of non-conforming product)
8.4 Analysis of Data (Analyse effectiveness and identify improvements)
8.5 Improvement (Corrective action, preventive action)

Figure 9.4 The main headings of ISO 13485 (Headings produced with permission of International Standards Organisation).

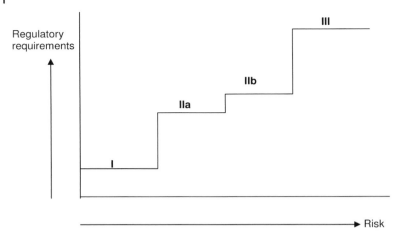

Figure 9.5 Relationship between risk and classification of medical devices.

to subject all devices to the same level of regulatory control. Instead, a tiered approach has been adopted that adjusts the level of regulatory oversight to the degree of potential hazard ascribed to the use of the device. This is illustrated in Figure 9.5.

Fundamental to this strategy is a procedure for classification of devices into one of four categories: classes I, IIa, IIb or III. Manufacturers must classify their devices according to criteria and rules set out in Annex IX to the directive, as amended. As a first step, the manufacturer must clearly define the intended use of the device in terms of: (i) degree of invasiveness; (ii) mode of action, whether active or passive device; (iii) the duration of contact with the patient; and (iv) impact on the body, local versus systemic effect.

- *Degree of invasiveness:* under invasiveness criteria, devices may be categorised as non-invasive, invasive through a body orifice (mouth, eye, ear, etc., or a permanent artificial opening such as for a colostomy bag), or surgically invasive. Note: Any device that is intended to penetrate the skin, such as a syringe needle, is considered to be a surgically invasive device.

- *Mode of action:* active devices depend on a source of electrical energy or other power, excluding gravity or the patient to perform their function.

- *Duration of contact:* devices are considered transient if intended for continuous use for less than 60 minutes, short term if intended for continuous use for not more than 30 days, and long term if intended for continuous use for more than 30 days. Devices that are intended to be introduced into the body or to replace an epithelial surface or the surface of the eye by a surgical procedure for more than 30 days are considered as implants.

- *Impact on the body:* the impact on the body may be defined as local if the effect is confined to the site around the area of application of the device, or systemic if the effects may spread throughout the body, as is the case when the circulatory or central nervous systems are involved.

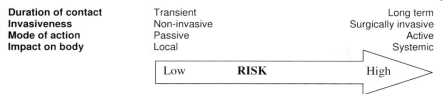

Duration of contact	Transient	Long term
Invasiveness	Non-invasive	Surgically invasive
Mode of action	Passive	Active
Impact on body	Local	Systemic

Figure 9.6 Impact of device characteristics on risk classification.

A general illustration of how these criteria impact on the risk classification is shown in Figure 9.6.

Having identified the characteristics of the device, the manufacturer must then apply the classification rules. There are 18 rules: four for non-invasive devices; four for invasive devices; four for active devices; and six special rules. These are shown in Figure 9.7. The device must be assigned the highest class that applies under the various rules. A decision chart illustrating the application of the classification rules for non invasive devices is shown in Figure 9.8. (The reader is referred to MEDDEV guide 2.4 for detailed guidance and examples of device classification; see Section 9.6.)

Manufacturers do not have to classify active implantable medical devices, but these are regulated as high-risk devices under the specific AIMD directive. Similarly, there is no requirement to classify IVDs under the IVD directive. Most IVDs are regulated as low-risk devices under this directive, except for tests that underpin the safety of blood and blood products (blood group, HIV and hepatitis tests), where additional specific requirements equating to a high-risk category apply.

9.3
Regulatory Strategy for Medical Devices in the US

The regulation of medical devices in the US was introduced with the Medical Device Amendments to FDC Act in 1976. These are supported by regulations published in 21 CFR parts 800-899. The Center for Devices and Radiological Health (CDRH) is the responsible section within the FDA for regulation of devices.

9.3.1
Classification of Devices

As in Europe, the classification of devices plays a central role in the regulation of devices in the US. However, there are a number of differences compared to the European approach. First of all, responsibility for classification rests with the FDA rather than with the manufacturers. When the regulations were introduced, the FDA mandated expert advisory panels (classification panels) to consider the different types of device that existed on the market at that time. The FDA provided classification questionnaires to act as guidelines for the panels when assessing the devices. In determining the safety and effectiveness of a device for purposes of classification, the classification panels had to consider the intended user (not part of EU criteria), the intended use/conditions of use, the risk-to-benefit ratio, and the reliability of the device. Proposals for classification of devices that could be grouped based on shared intended use and technical features were passed on to the FDA Commissioner, together with recommendations for exemptions or application/development of appropriate performance standards. After final consideration by the FDA, official

1. Non-invasive devices
Rule 1
All non-invasive devices are in Class I, unless one of the rules set out hereinafter applies.
Rule 2
All non-invasive devices intended for channelling or storing blood, body liquids or tissues, liquids or gases for the purpose of eventual infusion, administration or introduction into the body are in Class IIa:
- if they may be connected to an active medical device in Class IIa or a higher class,
- if they are intended for use for storing or channelling blood or other body liquids or for storing organs, parts of organs or body tissues,
in all other cases they are in Class I.
Rule 3
All non-invasive devices intended for modifying the biological or chemical composition of blood, other body liquids or other liquids intended for infusion into the body are in Class IIb, unless the treatment consists of filtration, centrifugation or exchanges of gas, heat, in which case they are in Class IIa.
Rule 4
All non-invasive devices which come into contact with injured skin:
- are in Class I if they are intended to be used as a mechanical barrier, for compression or for absorption of exudates,
- are in Class IIb if they are intended to be used principally with wounds which have breached the dermis and can only heal by secondary intent,
- are in Class IIa in all other cases, including devices principally intended to manage the micro-environment of a wound.

2. Invasive devices
Rule 5
All invasive devices with respect to body orifices, other than surgically invasive devices and which are not intended for connection to an active medical device or which are intended for connection to an active medical device in Class I:
- are in Class I if they are intended for transient use,
- are in Class IIa if they are intended for short-term use, except if they are used in the oral cavity as far as the pharynx, in an ear canal up to the ear drum or in a nasal cavity, in which case they are in Class I,
- are in Class IIb if they are intended for long-term use, except if they are used in the oral cavity as far as the pharynx, in an ear canal up to the ear drum or in a nasal cavity and are not liable to be absorbed by the mucous membrane, in which case they are in Class IIa.
All invasive devices with respect to body orifices, other than surgically invasive devices, intended for connection to an active medical device in Class IIa or a higher class, are in Class IIa.
Rule 6
All surgically invasive devices intended for transient use are in Class IIa unless they are:
- intended specifically to control, diagnose, monitor or correct a defect of the heart or of the central circulatory system through direct contact with these parts of the body, in which case they are in Class III,

Figure 9.7 The rules for classification of medical devices in the EU.

- reusable surgical instruments, in which case they are in Class I,
- intended specifically for use in direct contact with the central nervous system, in which case they are in Class III,
- intended to supply energy in the form of ionising radiation in which case they are in Class IIb,
- intended to have a biological effect or to be wholly or mainly absorbed in which case they are in Class IIb,
- intended to administer medicines by means of a delivery system, if this is done in a manner that is potentially hazardous taking account of the mode of application, in which case they are in Class IIb.

Rule 7

All surgically invasive devices intended for short-term use are in Class IIa unless they are intended:
- either specifically to control, diagnose, monitor or correct a defect of the heart or of the central circulatory system through direct contact with these parts of the body, in which case they are in Class III
- or specifically for use in direct contact with the central nervous system (CNS), in which case they are in Class III,
- or to supply energy in the form of ionizing radiation in which case they are in Class IIb,
- or to have a biological effect or to be wholly or mainly absorbed in which case they are in Class III,
- or to undergo chemical change in the body, except if the devices are placed in the teeth, or to administer medicines, in which case they are in Class IIb.

Rule 8

All implantable devices and long-term surgically invasive devices are in Class IIb unless they are intended:
- to be placed in the teeth, in which case they are in Class IIa,
- to be used in direct contact with the heart, the central circulatory system or the CNS, in which case they are in Class III,
- to have a biological effect or to be wholly or mainly absorbed, in which case they are in Class III,
- or to undergo chemical change in the body, except if the devices are placed in the teeth, or to administer medicines, in which case they are in Class III.

3. Additional rules applicable to active devices

Rule 9

All active therapeutic devices intended to administer or exchange energy are in Class IIa unless their characteristics are such that they may administer or exchange energy to or from the human body in a potentially hazardous way, taking account of the nature, the density and site of application of the energy, in which case they are in Class IIb.

All active devices intended to control or monitor the performance of active therapeutic devices in Class IIb, or intended directly to influence the performance of such devices are in Class IIb.

Rule 10

Active devices intended for diagnosis are in Class IIa:
- if they are intended to supply energy which will be absorbed by the human body, except for devices used to illuminate the patient's body, in the visible spectrum,

Figure 9.7 (*Continued*)

classifications and ancillary information relating to the regulation of the device types were published as regulations in 21 CFR Parts 862-892. The outcome of this process was that all medical devices, irrespective of their type, were classified into one of three classes.

- if they are intended to image in vivo distribution of radiopharmaceuticals,
- if they are intended to allow direct diagnosis or monitoring of vital physiological processes, unless they are specifically intended for monitoring of vital physiological parameters, where the nature of variations is such that it could result in immediate danger to the patient, for instance variations in cardiac performance, respiration, activity of CNS in which case they are in Class IIb.

Active devices intended to emit ionizing radiation and intended for diagnostic and therapeutic interventional radiology including devices which control or monitor such devices, or which directly influence their performance, are in Class IIb.

Rule 11

All active devices intended to administer and/or remove medicines, body liquids or other substances to or from the body are in Class IIa, unless this is done in a manner:
- that is potentially hazardous, taking account of the nature of the substances involved, of the part of the body concerned and of the mode of application in which case they are in Class IIb.

Rule 12

All other active devices are in Class I.

4. Special Rules

Rule 13

All devices incorporating, as an integral part, a substance which, if used separately, can be considered to be a medicinal product, as defined in Article 1 of Directive 2001/83/EC, and which is liable to act on the human body with action ancillary to that of the devices, are in Class III.

All devices incorporating, as an integral part, a human blood derivative are in Class III.

Rule 14

All devices used for contraception or the prevention of the transmission of sexually transmitted diseases are in Class IIb, unless they are implantable or long-term invasive devices, in which case they are in Class III.

Rule 15

All devices intended specifically to be used for disinfecting, cleaning, rinsing or, when appropriate, hydrating contact lenses are in Class IIb.

All devices intended specifically to be used for disinfecting medical devices are in Class IIa, unless they are specifically to be used for disinfecting invasive devices in which case they are in Class IIb.

This rule does not apply to products that are intended to clean medical devices other than contact lenses by means of physical action.

Rule 16

Devices specifically intended for recording of X-ray diagnostic images are in Class IIa.

Rule 17

All devices manufactured utilizing animal tissues or derivatives rendered non-viable are Class III except where such devices are intended to come into contact with intact skin only.

Rule 18

By derogation from other rules, blood bags are in Class IIb.

Figure 9.7 (*Continued*)

9.3.1.1 Class I Devices

Class I devices are considered low-risk devices and represent approximately 50% of all devices regulated by the FDA. A device is assigned to Class I if general controls are sufficient to provide reasonable assurance as to the safety and effectiveness of the device. Devices that are not life-supporting or life-sustaining, or are for a use which is

NON INVASIVE DEVICES

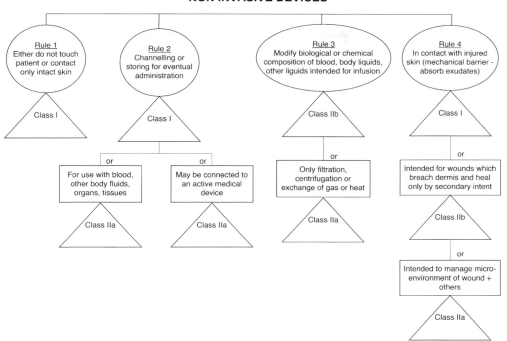

Figure 9.8 Decision chart for classification of non invasive devices (from MEDDEV Guide 2.4).

of substantial importance in preventing impairment of human health, and which do not present a potential unreasonable risk of illness of injury, may also be placed in Class I, even if there is insufficient information available to determine if additional controls are necessary to assure safety and effectiveness. The general controls to which Class I devices are subject under the FDC Act include Section 501 adulteration (contamination/cleanliness), Section 502 misbranding (false or misleading claims), Section 510 registration (establishment registration, inspection, pre-market notification), Section 516 banned devices, Section 518 notification and other remedies (administrative actions on defective devices), Section 519 records and reports (reporting of defects and adverse incidents) and Section 520 general provisions (good manufacturing practices as defined by the Quality System Regulation). In practice, most Class I devices are exempt from the pre-market notification requirements of Section 510k of the Act, unless they are life-supporting, life-sustaining or have a use that is of substantial importance in preventing impairment of human health. Some devices may also be exempt from compliance with the Quality System Regulation.

9.3.1.2 Class II Devices

Class II devices are viewed as moderate-risk devices, and represent the majority of the remaining devices. A device is assigned to Class II if there is sufficient information to establish special controls to augment the general controls and so provide reasonable assurance of the safety and effectiveness of the device. The special controls include the use of performance standards, the issuing of guidance documents and other measures as appropriate. The FDA themselves may develop performance standards which, following consultation, are published in the *Federal Register*. Alternatively, they can recognise existing standards developed by national or international bodies in similar fashion to the harmonised standards recognition procedure practised by the EU. All Class II devices are subject to Section 510k pre-market notification requirements.

9.3.1.3 Class III Devices

Class III devices are considered high risk and are usually technologically innovative devices. A device is assigned to Class III if it is considered that a combination of general and special controls is not sufficient to assure the safety and effectiveness of the device and it is life-supporting, life-sustaining, for use that is of substantial importance in preventing impairment of health or presents a potential unreasonable risk of illness or injury. The classification of implants or life-supporting or life-sustaining devices as other than Class III requires a positive justification for not doing so. All Class III devices must go through a pre-market approval process in order to establish the safety and effectiveness of the device.

9.3.2
Classification of New Devices

Manufacturers of new devices can determine the classification of their device by comparing its intended use and technological features to those of device types that have been already classified. The CDRH section of the FDA website contains a database of existing classifications that can easily be searched. Careful attention must be paid to the intended use in particular, as this can have a big bearing on how a device is regulated. For example, a human chorionic gonadotrophin (HCG) test system for the detection of pregnancy is regulated as a Class II device, whereas a HCG test system for any other purpose, such as detection or monitoring of certain types of cancer, is regulated as a Class III device. The trawl of device classifications will usually result in the identification of an existing device type that is "substantially" equivalent to the new device. Thus, the same classification can be applied. However, in cases of technologically innovative devices it may not be possible to identify an existing predicate device, and it will be viewed by default as a Class III device in such circumstances. However, a manufacturer of a new (or indeed an existing) Class III device may petition the FDA for a down-classification based on supporting objective scientific evidence as to its safety and effectiveness. The FDA itself is also pro-active in the re-assessment of Class III devices as market experience is gained as regards their performance, reliability and safety.

9.4
Development of Devices

The developments of drugs and devices share a common goal of producing products that are safe and effective. However, they differ in the approaches used for product development. Drugs rely heavily on research to identify new active substances, which are then subjected to extensive testing to establish their safety and effectiveness. The development of devices, on the other hand, relies on the application of technology to realise a product that meets an identified need. Thus, good design processes are central to the creation of devices that are capable of satisfying user needs and regulatory requirements.

Some form of quality system must be employed to ensure that devices satisfy the regulatory requirements for safety and effectiveness. At its most basic, a quality system may be applied to the final test and inspection of a finished product. However, significant product failure at this stage in the manufacturing process will be quite costly, as most of the labour, raw material and other manufacturing overheads will have been added to the product by this time. By applying a quality system to the complete manufacturing process, the risk and cost of product failures can be reduced. This focuses on ensuring that all raw materials are suitable and that manufacturing processes are robust and reliable. Thus, problems can be identified at an early stage, before significant manufacturing cost is incurred. Manufacturing quality assurance (QA) systems are essential to assure processes such as sterilisation or freeze-drying, where a strategy of a final inspection of all products cannot be used because of the destructive nature of the testing (breaking of seals). However, a manufacturing QA system is not focused to eliminate problems inherent in the design of the products. Unfortunately, many device failures on the market have been traced back to shortcomings in the initial design. The most effective way to address this is to employ a full QA system that covers both design and manufacture. By applying the design control elements of a full QA system, problems with the design can be identified and eliminated at the earliest stage, so that only reliable robust products that meet their intended use eventually reach the production stage. These basic principles are shown schematically in Figure 9.9.

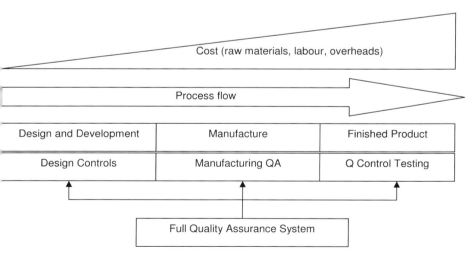

Figure 9.9 The principle of applying a full quality assurance (QA) system to minimise product failures and reduce costs.

9.4.1
Design Controls

Although, European regulations do not specifically require developers to adopt design controls, this is the preferred way to ensure that safe and effective devices are developed. Thus, European manufacturers will tend to follow the design control procedures that are set out in Section 7.3 of ISO 13485. Manufacturers of Class II and Class III devices for the US market are legally obliged to adopt the design controls set out in 21 CFR Section 820.30 of the Quality System Regulation (QSR). Most Class I devices are exempt from the design control elements of the QSR, except for devices automated with computer software and a number of specified devices such as surgical gloves. As illustrated in Figure 9.10, the design control elements of ISO13485 and the QSR are quite similar. The requirements for Design Transfer and Design History Files in the QSR are addressed under other headings in the ISO standard (4.2.4 Records, 7.3.1 Design & Development Planning). Thus, although there may be variations in emphasis between the two standards, a developer does not have to do anything significantly different to comply with either standard. As a basic step, manufactures will usually establish an overall design control procedure that describes how they address the elements of the standard when developing their devices.

9.4.2
Design and Development Planning

Manufacturers are required to establish and maintain plans that describe the design and development activities and define the responsibilities and authorities for completion of the design – that is, what must be done, and who must do it. A simplified outline of the stages that a development project may go through is shown in Figure 9.11, although the activities can vary considerably depending on the type and complexity of the device involved. Although the bulk of the activity may be confined to a core (R&D) development team, the project will invariably involve interaction with other groups within the organisation, such as marketing for identifying customers' needs, manufacturing and quality control for design transfer

ISO 13485 **7.3 Design and Development**	21 CFR **820.30 Design Controls**
7.3.1 Design and Development Planning	b) Design and Development Planning
7.3.2 Design and Development Inputs	c) Design Input
7.3.3 Design and Development Outputs	d) Design Output
7.3.4 Design and Development Review	e) Design Review
7.3.5 Design and Development Verification	f) Design Verification
7.3.6 Design and Development Validation	g) Design Validation
7.3.7 Control of Design and Development Changes	h) Design Transfer
	i) Design Changes
	j) Design History File

Figure 9.10 Comparison of headings of the design control elements of ISO13485 and the Quality System Regulation (QSR).

Stage	Description
Design Input	Identifying user needs, regulatory requirements, applicable standards
Design Development	Experiments and trials to create a prototype device and supporting documentation
Design Verification	Testing of the prototype to see if it meets design input requirements
Design Transfer	Activities to ensure that device can be manufactured under production conditions
Design Validation	Testing of the device under intended use conditions, field trials, clinical trials
Design Release	Formal release of device for manufacturing, regulatory marketing authorisations

Figure 9.11 Possible design and development stages for a simple device.

requirements and regulatory affairs for regulatory matters. These technical interfaces must be clearly defined in the plans. The plans should also identify key milestones and design reviews checkpoints, and for effective coordination and management will normally contain estimates of the time required for each task. *Gantt charts* are frequently used as a planning tool for presenting an overview of the information, as this enables the display of the task, the schedule, the progress, the responsibility and the milestones on a single chart. This approach is shown in Figure 9.12, where the tasks from Figure 9.11 are transformed into a hypothetical Gantt chart.

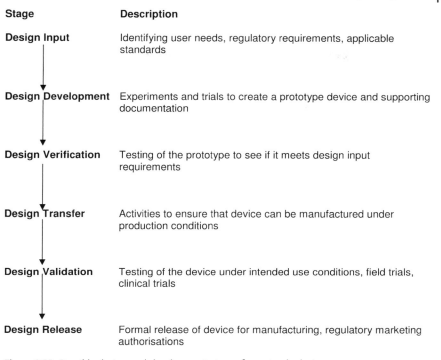

Time (Months)	1	2	3	4	5	6	7	8	9	10	11	12	13	14										
Task																								
Design Input	■	■	R&D, RA/QA, Marketing, Manufacturing																					
Design Review		◇	R&D, RA, Marketing, Production																					
Design Development			■	■	■	■	R&D																	
Design Verification								■	R&D															
Design Review									◇	R&D, RA/QA, Marketing, Manufacturing														
Design Transfer									▨	R&D, Production, QC														
Design Validation									▨	▨	R&D, RA, Field Site												
DesignReview											◇	R&D, RA/QA, Marketing, Manufacturing												
Design Release										▨	▨	R&D, RA, Document Contol											
Design Review												◇	R&D, RA/QA, Marketing, Manufacturing											

■ = completed   ▨ = planned   ◇ = milestone

Figure 9.12 Gantt chart for the tasks outlined in Figure 9.11.

9.4.3
Design Input

Design input is the process of collecting the various requirements that a device must meet, and translating them into a set of performance and other specifications that define the design task in as much detail as is feasible at that point. This should result in an authorised design input specification document. Inputs for such a document can come from many sources, and will normally include: user requirements as captured by marketing; regulatory requirements as advised by regulatory affairs; existing standards; manufacturing requirements or limitations; and the conclusions from a risk analysis conducted at this stage.

To illustrate the process of how basic requirements can be translated into an input specification, consider a simple glucose test. The user requirement from the market might be to develop a test for serum or urine that can be used in emergency situations, such as can arise from an insulin-induced diabetic coma. Based on a knowledge of the glucose levels that can be found in serum or urine, a required measuring range for the assay can be defined. The fact that samples in an emergency could be from non-fasting individuals would mean that the assay should not be subject to interference from turbid sera that might be encountered if the subject had recently consumed a meal high in lipid content. Other goals for accuracy and precision could be dictated by the existence of reference standards for calibration and performance standards for laboratory assays. A consultation of US device classifications would identify it as a Class II device, and thus determine the regulatory approach and the need to select a suitable reference product for the demonstration of substantial equivalence. The final outcome would be a design input document that contains quite detailed specifications as regards the required features of the device, and which can be objectively assessed once the device is developed.

Because of the expectation for a reasonably detailed design input specification, design controls are difficult to apply to initial research, where the feasibility of product concepts may be investigated. Such initial feasibility studies may be excluded, but once there is a clear design brief as to the product that needs to be developed, then the design controls must be applied. For complex projects with multiple design steps it may not be possible to specify all of the design inputs at the outset. Instead, the design output from a completed stage may form the basis for the design input for the next stage, so that design input evolves as the project progresses.

9.4.4
Design Output

Design output relates to what comes out at the end of a successful design process or stage. Rather than the physical device, design output usually takes the form of a set of documents or other materials describing how the device is manufactured and used. These could be presented as drawings, recipes, or procedures for manufacture and test of the device, specifications for raw materials, packaging, labelling or acceptance criteria, and any other element necessary to ensure the proper functioning and use of the device. The standards require that all such outputs are maintained, reviewed and authorised according to defined procedures.

9.4.5
Design Verification and Design Validation

Design verification and design validation are both activities related to investigations of the performance of the developed device. However, there are important features that distinguish them from each other:

- *Design verification* is focussed on providing test data or other objective evidence to show that devices or device elements which are the result of the design output meet the design input specifications. Typically, design verification is carried out in-house on pre-production batches. Continuing the example of the glucose test, design verification would involve using a set of glucose standards to verify that the reagent produced according to the recipe defined in the design output, met the required linear measuring range specified in the design input. Similarly, artificially turbid samples may be used to investigate lipaemic interference. Further checks would be made to verify that all the other individual design input specifications were achieved.

- *Design validation*, on the other hand, is focussed on assessing if the device in its totality meets the user's needs and intended uses and functions correctly under the intended use conditions. Thus, evaluation will usually be carried out at field sites/β-sites (hospitals) or, at a minimum, under simulated use conditions. In the case of high-risk devices this will normally involve actual clinical studies. Validation should be performed using initial production lots of the device or their equivalents. An advantage of using production lots is that the validation testing will provide confirmation that the design transfer has been effective. The devices should be provided with representative packaging, labels and instructions for use so as to identify any difficulty, ambiguity, or misunderstanding arising from how the device is packaged and presented. For the glucose test, the measuring range would be confirmed by assaying real samples containing different levels of glucose and different degrees of turbidity and comparing results to a reference method.

A schematic representation of the relationship between design verification and design validation and their interaction with other design activities is shown in Figure 9.13.

9.4.6
Design Review

The standards require that formal design review meetings be conducted at appropriate stages in the process. The meetings should be attended by representatives of personnel responsible for the design stage, together with other experts as appropriate

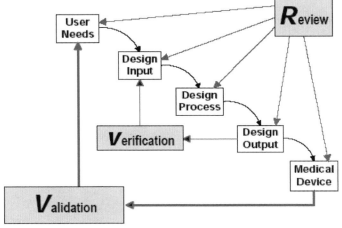

Figure 9.13 Application of design controls to the waterfall design process. (Figure reproduced with permission of the Minister of Publik Works and Government Services Canada, 2008).

and, in the case of the QSR, one independent reviewer with no responsibility for the stage under review. The number of reviews will vary depending on the complexity of the project, but typically reviews may be conducted at the end of design input, design verification, design validation and final product release. Records of the results of such meetings must be documented and maintained.

9.4.7
Risk Analysis

Risk analysis is usually conducted as part of the design review meetings. The EU directives require that the general principles of risk management in terms of elimination, reduction, protection or warning be applied to the design and manufacture of devices. The ISO standard also calls for risk management as part of the overall product realisation strategy, while the QSR specifically refers to risk analysis under design validation. Thus, manufacturers need to adopt measures that can identify and quantify risks. Two common techniques are used for identifying risk, Fault Tree Analysis (FTA) and Failure Mode Effect Analysis (FMEA). In the FMEA approach, consideration is given to what would be the end-consequence of failure of a component or element of the device. FTA takes the reverse view, and considers a fault in the devices and asks what are the different causes that could result in the failure. Having identified the risk, manufacturers need some way of quantifying the risk as objectively as possible. In doing so, they may develop a scoring matrix based on:

- the probability of the failure occurring;
- the severity of the consequential hazard; and
- the ability to detect a failure.

Finally, while the first preference would be to eliminate all risks, a risk management strategy should be established that identifies what level of risk can be tolerated and controlled.

9.4.8
Design Changes

All design changes need to be agreed and controlled. For example, if it proved impossible to achieve one of the original design inputs, but an adequate device could still be created in it absence, then a revised design input would have to be agreed and authorised by the approvers of the original document. Consideration must also be given to the impact of the design change and the need to repeat verification or validation studies.

9.5
Chapter Review

Because of the extensive range of devices subject to regulation, the classification of devices is a fundamental element of the regulatory strategy in both Europe and the US. This provides the basis for the application of a tiered level of regulatory requirements. Unfortunately, classification procedures and regulatory requirements remain to be harmonised between both jurisdictions, although both regions promote the use of standards to ensure that specific types of device are safe and effective. The application of design controls as defined in ISO standards or the US Quality System Regulation is central to the design and development of many devices.

9.6
Further Reading

- Standardisation
  www.newapproach.org.
  www.cen.eu.
  www.cenelec.org.
  www.iso.org.
  www.iec.ch.
- Harmonised European Standards
  http://ec.europa.eu/enterprise/newapproach/standardization/harmstds/reflist/meddevic.html.
- EU Device Classification
  MEDDEV Guide 2.4
  http://ec.europa.eu/enterprise/medical_devices/meddev/index.htm.
- EU Directive 93/42/EEC
  http://ec.europa.eu/enterprise/medical_devices/legislation_en.htm.

- US Device Classification Database
www.accessdata.fda.gov/scripts/cdrh/cfdocs/cfpcd/classifiaction.cfm.
- US Regulations & Guidance
Design control regulations 21 CFR 820.30
FDA Design Control Guidance for Medical Device Manufacturers
www.fda.gov.

10
Authorisation of Medical Devices

10.1
Chapter Introduction

The previous chapter outlined how device classification and the use of standards provide the basis for effective regulation of medical devices, with particular focus on the application of design control standards to the development of devices. In this chapter we look at the process for evaluation and authorisation of devices, and see how the regulatory requirements vary depending on the perceived risk of the device as indicated by its classification. It will be noted that there is considerable variation between the approaches adopted in Europe and the US and that, compared to drugs, practical harmonisation of requirements still remains to be adopted.

10.2
Evaluation of Medical Devices in Europe

As a general rule, clinical data are required as evidence to support conformity with the requirements of the Active Implantable Medical Devices (AIMD) and the Medical Device (MD) directives with regards to safety and effectiveness under the normal conditions of use, evaluation of undesirable side effects, and the acceptability of the benefit/risk ratio. Risk analysis should be used to establish key objectives that need to be addressed by clinical data, or alternatively to justify why clinical data are not required (mainly for Class I devices). The risk analysis process should help the manufacturer to identify known (or reasonably foreseeable) hazards associated with the use of the device, and decide how best to investigate and estimate the risks associated with each hazard. The clinical data should then be used to establish the safety and effectiveness of the device under the intended use conditions, and to demonstrate that any of the residual risks are acceptable, when weighed against the benefits derived from use of the device.

The provision of clinical data in support of conformity with the requirements of the directives does not necessarily mean that the device has to be subjected to actual clinical investigations. Instead, the manufacturer may present critical evaluations of

Medical Product Regulatory Affairs. John J. Tobin and Gary Walsh
Copyright © 2008 WILEY-VCH Verlag GmbH & Co. KGaA, Weinheim
ISBN: 978-3-527-31877-3

relevant scientific literature or existing clinical investigation data or other source material as clinical data. This can be based on studies of an equivalent device, provided that the manufacturer can establish such equivalence in terms of clinical use and technical or biological parameters, with particular emphasis on performance, principles of operation and materials. Studies should have validity as regards scientific methodology and the capability of the investigators involved. The data should be presented in a report that identifies its source, and critically evaluates its relevance and the conclusions that can be drawn. If the manufacturer is unable to address satisfactorily all of the safety and effectiveness issues by reference to existing data alone, then specific clinical investigations will have to be conducted. In the case of implantable devices and devices in Class III, clinical investigations will be required as the norm, unless the manufacturer is able to provide positive justification as to why it is sufficient to rely on existing clinical data.

Studies to investigate the safety and effectiveness of In Vitro Diagnostic (IVD) medical devices under intended use conditions are conducted as performance evaluations. They are considered to present less risk than clinical investigations since, by their nature, studies involving IVDs cannot have any direct impact on the health and safety of trial subjects.

10.2.1
Clinical Investigations

Clinical investigations do not come within the scope of the Good Clinical Practice Directives (2001/20/EC and 2005/28/EC), which only regulate trials involving human medicinal products (pharmaceuticals). Instead, basic requirements on the conduct of clinical investigations are stated in Annex 7 to the AIMD directive and Annex X to the MD directive. The main provisions are outlined in Figure 10.1. Manufacturers may adopt harmonised ISO standard 14155 on "Clinical investigation of medical devices for human subjects" to demonstrate compliance with the requirements of the directive. In practice, and in light of the 2007 update to the device directives, the requirements for conducting clinical investigations are quite similar to those for clinical trials, as outlined in Chapter 6. Thus, trials will require criteria for participant selection, prior informed consent forms for subjects, a favourable opinion from an ethics committee, a clinical investigation plan, an investigator's brochure, qualified investigators, case report forms, adverse event reporting and study reports. The devices used in the investigation must conform to the essential requirements with the exception of those aspects that are the subject of the study. Thus, packaging, labelling and instructions for use should be close to the final version. Device labels should carry the statement "exclusively for clinical investigations".

Before initiating a clinical investigation, the manufacturer (or his/her authorised representative) must prepare a statement containing the information outlined in Figure 10.2, and notify the relevant Competent Authorities of the proposed study. The manufacturer must also keep available for inspection by the Competent Authorities the information outlined in Figure 10.3. Unlike for clinical trials, a standard application form has not been developed. Instead, appropriate application forms are

Ethical Considerations
Investigations must be conducted in accordance with the ethical principles contained in the Helsinki Declaration.

Clinical Investigation Plan
Investigations must be conducted on the basis of an appropriate investigational plan that addresses all the key aims of the study.

Intended Use Conditions
Investigation should be performed in circumstances similar to the intended use conditions.

Adverse Event Reporting
All adverse events that cause or have the potential to cause death or serious injury must be reported to the Competent Authorities of the Member States where the investigations are being conducted.

Investigator
The investigations must be performed under the responsibility of a medical practitioner or other authorised qualified person (e.g. dentist).

Investigation report
A written report must be prepared summarising the investigational plan, methods, and results with appropriate discussions and conclusions. The investigator must sign the report.

Figure 10.1 Key requirements for the conduct of clinical investigations with medical devices.

- Data allowing identification of the device in question

- An investigation plan, stating in particular the purpose, scientific, technical or medical grounds, scope and number of devices concerned

- The investigator's brochure, the case report forms, the confirmation of insurance of subjects and the documents used to obtain informed consent

- The opinion of the ethics committee concerned and details of the aspects covered by its opinion

- The name of the medical practitioner or other authorized person and of the institution responsible for the investigations

- The place, starting date and scheduled duration for the investigations

- A statement that the device in question conforms to the essential requirements apart from the aspects covered by the investigations and that, with regard to these aspects, every precaution has been taken to protect the health and safety of the patient.

Figure 10.2 Information to be included in a clinical investigation statement.

- A general description of the product and its intended use

- Design drawings, methods of manufacture envisaged, in particular as regards sterilization, and diagrams of components, sub-assemblies, circuits, etc.

- The descriptions and explanations necessary to understand the above-mentioned drawings and diagrams and the operation of the product

- The results of the risk analysis and a list of harmonised standards applied in full or in part, and descriptions of the solutions adopted to meet the essential requirements of this Directive if the harmonised standards have not been applied

- A declaration stating whether or not the device incorporates, as an integral part, a substance or human blood derivative, and the data on the tests conducted in this connection which are required to assess the safety, quality and usefulness of that substance or human blood derivative, taking account of the intended purpose of the device

- A statement indicating whether or not the device is manufactured utilising tissues of animal origin as referred to in Directive 2003/32/EC and the risk management measures in this connection which have been applied to reduce the risk of infection

- The results of the design calculations, and of the inspections and technical tests carried out, etc.

Figure 10.3 Information relating to clinical investigations that must be maintained available to Competent Authorities for inspection.

usually available from individual Competent Authorities. In the case of investigations involving Class III, active implantable, and implantable or long- term invasive devices of Class IIa or Class IIb, the Competent Authority is allowed 60 days to review and authorise the study. For other lower-risk devices, the Competent Authorities may immediately issue a letter of no objection, provided that a relevant ethics committee has sanctioned the investigation. Data on ongoing clinical investigations will be maintained available to all Competent Authorities via the European Databank for medical devices (see Section 10. 4. 10 for information on the European Databank).

10.2.2
Performance Evaluations

Although most performance evaluations of IVDs can be conducted on samples taken from subjects during the course of normal clinical practices, ethical considerations require that informed consent be obtained for the use of these samples for research purposes. Many institutions use a general consent form when taking samples from patients. Where specific sampling is required, studies should be planned so as to minimise the discomfort or impact on subjects. Results obtained with evaluation devices should not be relied on to decide the clinical care of subjects. Before commencing an evaluation, the manufacturer must draw up a statement containing the

- Data allowing identification of the device in question

- An evaluation plan stating in particular the purpose, scientific, technical or medical grounds, scope of the evaluation and number of devices concerned

- The list of laboratories or other institutions taking part in the evaluation study

- The starting date and scheduled duration for the evaluations and, in the case of devices for self-testing, the location and number of lay persons involved

- A statement that the device in question conforms to the requirements of the Directive, apart from the aspects covered by the evaluation and apart from those specifically itemised in the statement, and that every precaution has been taken to protect the health and safety of the patient, user and other persons

Figure 10.4 The statement drawn up in advance of an IVD Performance Evaluation.

information outlined in Figure 10.4. There are no specific legal requirements for the conduct of performance evaluations, but manufacturers may adopt European standard EN 13612 on 'Performance evaluation of in vitro diagnostic medical devices' as best practice. This requires that manufacturers appoint a study coordinator, establish agreements with investigators and ensure that they are adequately informed and have appropriate resources, and prepare an evaluation plan and a final study report.

10.3
Evaluation of Medical Devices in the US

The conduct of studies of medical devices in the US that have not been cleared/approved by the FDA is regulated via Investigational Device Exemption (IDE) regulations set out in 21 CFR Part 812. Considering the type of device and the level of associated risk involved, investigations may be conducted as IDE-exempted studies, Abbreviated requirement studies, or studies subject to full IDE requirements.

10.3.1
IDE-Exempted Investigations

Exempted investigations may be conducted where the studies pose negligible risk to subjects. These include:

- Investigations of devices that have already received FDA-clearance due to the fact that they were already on the market before medical devices were regulated or were shown to be substantially equivalent to one.
- Diagnostic devices that are non-invasive, do not require significantly risky sampling procedures and do not introduce energy (e.g. X-rays, ultrasound) into the subject. The study results may not be relied on for diagnosis.

- Devices for veterinary use or intended for research on or with laboratory animals.
- Custom devices, provided that the studies are not being used to establish safety and effectiveness for commercial distribution.

While there is no FDA involvement in such investigations, ethical considerations and institutional regulations would generally dictate that informed consent and the approval of the local institutional review board be obtained. However, due to the negligible risks involved these are likely to be a formality and will not involve extensive review or scrutiny.

10.3.2
Abbreviated Requirement Investigations

Many investigations that do not involve highly invasive devices, risky procedures and/or frail patients can be conducted under abbreviated requirements applicable to devices that are not considered significant risk devices. The sponsor does not have to submit an IDE application to the FDA. Instead, the review and approval of such studies is the responsibility of Institutional Review Boards. However, where there is any doubt, it is good practice to obtain advance confirmation from the FDA that the device can be designated as a non-significant risk device. The investigations are subject to some of the measures associated with a standard IDE investigations, such as study monitoring by the sponsor, recording and reporting of adverse events, and the prohibition on promotion or advertising. A comparison of the differences in requirements between the investigation of a non-significant risk device and a significant risk device is shown in Table 10.1.

10.3.3
IDE Investigations

Investigations involving significant risk devices require prior approval by the FDA. A device is considered a significant risk if it presents a potential for serious risk to the health, safety, or welfare of a subject. Implants, devices that are life-supporting or sustaining and devices that are of substantial importance in diagnosing, treating or mitigating disease or otherwise preventing health impairment, generally fall into this category. An IDE application must be submitted to the FDA Center for Devices and Radiological Health (CDRH), accompanied by the information outlined in Figure 10.5. The required content of an Investigational Plan is shown in Figure 10.6. The FDA are allowed 30 days either to review and approve or reject the application. Changes to an approved IDE must be submitted to the FDA for approval as a supplemental IDE, unless the changes are emergency changes necessitated to preserve life or well-being, or developmental changes and alterations to protocols that do not significantly change the validity or safety of the investigations. These should be notified to the FDA, 5 days in advance of developmental or protocol amendments and 5 days after an emergency change. An overview of the procedures available for the investigation of devices is shown in Figure 10.7.

Table 10.1 Comparison of requirements for investigations involving non-significant risk versus significant risk devices.

Requirement	NSR S	NSR I	SR S	SR I
Submit an IDE application to FDA			✓	
Report adverse effects to sponsor		✓		✓
Report adverse effects to IRB	✓	✓	✓	✓
Report adverse effects to FDA			✓	
Report withdrawal of IRB approval to sponsor		✓		✓
Submit progress reports to sponsor, monitor and IRB		✓		✓
Report deviations from investigational plan to sponsor and IRB				✓
Obtain document and retain records of informed consent		✓		✓
Report use of device without informed consent to sponsor and IRB		✓		✓
Compile records of adverse effects and complaints	✓		✓	
Maintain all records of correspondence			✓	✓
Maintain records of device shipment use and disposal			✓	✓
Document day and time of use of each device				✓
Maintain signed investigator agreements			✓	✓
Investigator list to FDA every 6 months			✓	
Progress reports to IRB at least yearly	✓		✓	
Yearly progress reports to FDA			✓	
Final study report to FDA			✓	
Final study report to IRBs	✓		✓	
Monitor study and ensure compliance with protocol	✓		✓	
Notify IRBs if an investigational device has been recalled	✓		✓	
Comply with promotion and advertising prohibitions	✓	✓	✓	✓
Comply with investigational device labelling requirements	✓		✓	

NSR = non-significant risk device investigations; SR = significant risk device investigations; S = sponsor; I = investigator.
Requirement differences are in bold.

10.3.4
Labelling of Devices for Investigational Use

Devices intended for clinical investigations must bear the following statement on the labelling: "CAUTION – Investigational device. Limited by Federal (or United States) law to investigational use."

Devices intended solely for research on or with laboratory animals should be labelled with: "CAUTION – Device for investigational use in laboratory animals or other tests that do not involve human subjects."

In-vitro diagnostic reagents for field trial studies should bear the statement "For Investigational Use Only. The performance characteristics of this product have not been established."

Investigational Plan
Reports of previous investigations and a summary of the proposed investigational plan or the plan itself.

Methods, Facilities, Controls
A description of the methods, facilities, and controls used for the manufacture, processing, packing, storage, and, where appropriate, installation of the device

Investigators
A list of investigators and an example of the agreement that they have entered into to comply with their obligations

Certification of Investigator Agreements
Certification that the list of investigators is complete, that they have signed the agreement and new investigators must also sign the agreement before participating

IRB Information
A list of names and address of chairpersons of Institutional Review Boards (IRBs) that have been asked to review the investigation

Other Institutions
The names and addresses of other institutions that may be involved in part of the investigation but are not identified under the IRB listing e.g. institutions that may perform supporting analytical testing

Device Charges
Proposed financial charges for devices that will not be provided free

Environmental Assessment
Certificate of exclusion from environmental assessment or assessment report as applicable

Labelling of devices

Figure 10.5 Documentation to be provided with an IDE application.

If the reagent is intended for an earlier stage laboratory investigation, then the label should state "For Research Use Only. Not for use in diagnostic procedures."

10.4
Placing of Devices on the Market in the EU

Before a device can be legally placed on the market in Europe, it must go through an appropriate conformity assessment process to establish that it meets all the essential requirements of the applicable directive. This enables the manufacturer to make a formal declaration of conformity and apply the CE mark of conformity to the device. The manufacturer, or an authorised representative (EC Rep), who takes on the responsibilities imposed by the directives on behalf of the manufacturer, must be established in the EU.

Purpose
Device description and intended use and objectives and duration of investigation

Protocol
Methodology to be used to conduct and analyse the study

Risk analysis
Risks identified from risk analysis, measures to minimise and control risks and information on subject selection

Description of the device
A description of each significant component/ingredient of the device and its function and how it might change during the investigation

Monitoring Procedures
A description of the procedures that will be used to monitor the study

Labelling

Consent materials
Copies of forms and informational materials used to obtain informed consent from subjects

IRB Information
Lists of names and addresses of chairpersons of IRBs that will review the investigations

Other institutions
A list of other institutions that may be involved in part of the investigation without IRB oversight

Additional Records and Reports
A description of any additional records or reports that may be maintained above those legally required

Figure 10.6 Required content of an Investigational Plan.

10.4.1
Conformity Assessment Procedures

Different conformity assessment options are available, depending on the type of device and the level of associated risk. For low-risk devices, the manufacturer can make a declaration of conformity based solely on self-assessment, without the need for the involvement of a Notified Body. For all other devices Notified Bodies are required to perform one or more of the tasks outlined in Table 10.2. HIV and hepatitis tests and blood grouping tests represent the highest risk devices, as they are critical to ensuring the safety of blood and blood products. For example, a defective HIV test device could result in widespread infection in an unsuspecting population, whereas the detrimental effects of an AIMD or a Class III device failure will just be confined to the individuals treated by the device. At this end of the risk spectrum, Notified Bodies are required to verify the applied quality system, the specific device design, and the

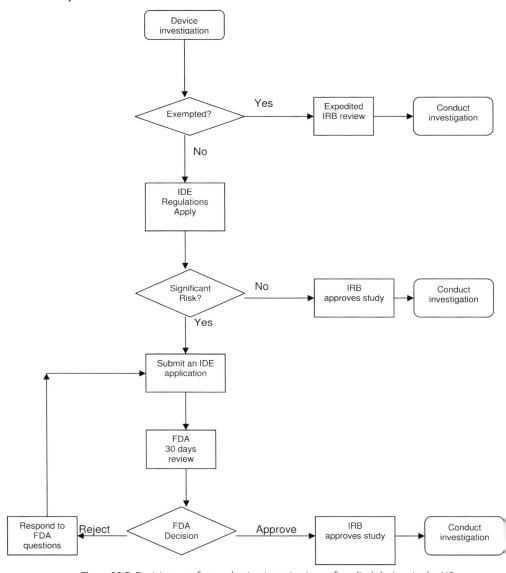

Figure 10.7 Decision tree for conducting investigations of medical devices in the US.

final testing of individual batches. The various options that may be applied to different categories of device are summarised in Figure 10.8.

The purpose of all the conformity assessment procedures is to ensure that the device as designed and as subsequently manufactured will meet the essential requirements. For most devices where Notified Body involvement is required, the manufacturer may choose between the strategies of relying on the application of appropriate quality systems or independent testing to demonstrate this.

Table 10.2 Certification activities of Notified Bodies.

Quality Systems Certification	Device Certification

Full QA System Certification
  Certification based on auditing against
  a quality system standard for design,
  production & final inspection (e.g. ISO 13485)
	EC Design-examination Certification
	  Certification based on examination
	  of the design dossier versus
	  standards/essential requirements
	  (paperwork review)

Production QA System Certification
  Certification based on auditing against
  a quality system standard for production &
  final inspection (e.g. ISO 13485 - design control)
	EC Type-examination Certification
	  Certification of design based on
	  testing of physical samples versus
	  standards/essential requirements

Product QA System Certification
  Certification based on auditing against a quality
  system standard for final inspection (e.g. EN 46003)
	EC Verification Certification
	  Certification of individual produc-
	  tion batches based on testing of all
	  or samples of the batch versus
	  standards/acceptance criteria

10.4.2
Full Quality Assurance

Conformance based on a full quality assurance (QA) model is the preferred option for high-risk devices. The manufacturer must adopt a QA system for the design, manufacture and final inspection of the products (ISO 13485). He/she must then lodge an application with a Notified Body to have the quality system audited and the particular design dossier examined. (Note: examination of the specific design dossier is not required for Class II devices.) The application must be accompanied by documentation describing the quality system and data enabling assessment of the specific design. The Notified Body will then review the documentation and carry out a site audit of the quality system. If satisfactory, the Notified Body may certify the quality system and issue an EC design-examination certificate for the design. The manufacturer must inform the Notified Body of any plans to substantially modify the quality system, and also obtain a further approval for any design changes that could affect the conformity of the design. The Notified Body must periodically audit the quality system to ensure that compliance is maintained.

10.4.3
EC Type-Examination

The EC type-examination is the process that the manufacturer may use to obtain independent verification that a design conforms to essential requirements, when a certified QA system has not been applied to the design process. The manufacturer must submit an application to the notified body, accompanied by documentation on the device design and physical samples of the device. The Notified Body examines the

Active Implantable Medical Devices Directive 90/385/EEC	Medical Devices Directive 93/42/EEC	In Vitro Diagnostic Medical Devices Directive 98/79/EC
	Class III	**List A** (blood group, HIV, Hepatitis)
Full QA & EC Design-examination (Annex II)	Full QA & EC Design-examination (Annex II)	Full QA & EC Design-examination (Annex IV) + Independent batch verification
Or	Or	Or
EC Type-examination (Annex III) + { EC Verification (Annex IV) or Production QA (Annex V) }	EC Type-examination (Annex III) + { EC Verification (Annex IV) or Production QA (Annex V) }	EC Type-examination (Annex V) + Production QA (Annex VII) + Independent batch verification
	Class IIb	**List B**
	Full QA (Annex II) Minus EC Design-examination	Full QA (Annex IV) Minus EC-design examination
	Or	Or
	EC Type-examination (Annex III) + { EC Verification (Annex IV) or Production QA (Annex V) or Product QA (Annex VI) }	EC Type-examination (Annex V) + { EC Verification (Annex VI) or Production QA (Annex VII) }
	Class IIa	**Self-Testing Devices**
	Full QA (Annex II) Minus EC Design-examination	EC Self Declaration (Annex III) + EC Design-examination
	Or	
	EC Self Declaration (Annex VII) + { EC Verification (Annex IV) or Production QA (Annex V) or Product QA (Annex VI) }	
	Class III	**Other IVDs**
	EC Self Declaration (Annex VII)	EC Self Declaration (Annex III)

(Risk increases from bottom to top, indicated by upward arrow on the right)

Figure 10.8 Conformity assessment options available to manufacturers for different types of device

documentation, tests the device samples and, if the design conforms to the require-ments, issues an EC type-examination certificate. This can be combined with a certified production QA system (manufacture and final inspection) or EC Verification to achieve a level of assurance approaching that of a full QA system.

10.4.4
Production Quality Assurance

Production QA is the process whereby the manufacturer adopts a quality system for the manufacture and final inspection of a device. ISO 13485 may also be used as the model in these circumstances, as the design and development elements may be omitted from this standard. The manufacturer must lodge an application with a Notified Body to have his/her system examined. The application must be accompanied by documentation on the quality system and information on any relevant EC type-examined devices.

10.4.5
EC Verification

EC verification provides an alternative to the model of establishing a certified production QA system. Independent testing of either all devices, or a statistically representative sample of each batch, is conducted by or on behalf of the Notified Body, which then issues a certificate of conformity for the tests conducted. This is not a popular option due to the costs involved. The procedure is not capable of providing adequate assurance as to the sterility of devices. Instead, an assurance of sterility must be based on the application of a production QA system to the sterilisation process.

10.4.6
Product QA

Product QA is the process whereby the manufacturer applies to a Notified Body to have his/her quality system for final inspection audited and certified. EN 46003 may be used as the harmonised standard for this model.

10.4.7
EC (Self) Declaration of Conformity

For low-risk devices the manufacturer may make a declaration of conformity with the essential requirements on the basis of a self-assessment, without any input from a Notified Body. In the case of Class I devices that are sterile or have a measuring function, Notified Body involvement is required, to provide assurance as to the effectiveness of the sterilisation process or the meteorological traceability of values as appropriate.

10.4.8
Technical Documentation

Assembling technical documentation that provides evidence of conformity with the essential requirements is a fundamental part of all conformity assessment procedures. The appropriate essential requirements' annex may be used as a checklist, against which documents demonstrating the adopted solutions may be identified and referenced. This documentation would include, performance data, procedures, relevant standards and labelling. Records of reports or certifications by a Notified Body and the manufacturer's declaration of conformity will also form part of the technical documentation. If technical documentation does not have to be submitted to a Notified Body, then it is simpler to cross-reference to existing files rather than assembling it into a single file. For example, data relating to the design could just be cross-referenced to the existing development file/design history file, while information on the quality system manufacturing, or labelling could be referenced to appropriate files. This facilitates keeping the file up to date, because for example if a package insert was revised due to the addition of new languages, the master labelling files would just have to be updated, without having to also update a "central" technical file. Thus, the master technical file may just consist of a cross-referencing index based on the essential requirements, copies of test certificates issued by Notified Bodies, and the declaration of conformity. The technical documentation must be maintained at the disposal of the Competent Authorities for a period of 5 years after manufacture of the last product. Requests from Competent Authorities to view the technical documentation will usually be triggered by problems in the market.

10.4.9
Labelling Requirements

The essential requirements provide reasonable detail on the requirements for labelling of the different categories of medical devices. These are supplemented by the harmonised standards outlined in Table 10.3. Instructions for use must be supplied with all devices, except for Class I and Class IIa devices which can be used safely without accompanying guidance. Information technology solutions may be used to provide this information, particularly in the case of *in-vitro* diagnostic reagents for professional use. In general, the labelling should focus on ensuring that the device is used safely and properly and, where appropriate, should address aspects such as intended use, handling and storage, warnings and precautions, principles of the device and operating instructions, expected performances and limitations. Labelling must be in the national languages of the Member States, with

Table 10.3 Harmonised standards for labelling of medical devices.

EN 375	Information supplied by the manufacturer with *in-vitro* diagnostic reagents for professional use
EN 376	Information supplied by the manufacturer with *in-vitro* diagnostic reagents for self-testing
EN 591	Instructions for use for *in-vitro* diagnostic instruments for professional use
EN 592	Instructions for use for *in-vitro* diagnostic instruments for self-testing
EN 980	Graphical symbols for use in the labelling of medical devices
EN 1041	Information supplied by the manufacturer with medical devices

Symbol	Meaning	Symbol	Meaning
⊗	Do Not Reuse	STERILE EO	Sterile + method of sterilisation
⧖	Use by	∮	Temperature limitation
LOT	Batch number or lot number	**SN**	Serial Number
ᴍ	Date of manufacture	**REF**	Catalogue number
⚠	Attention, see instructions for use	🔲i	Consult instructions for use
▲	Manufacturer	EC REP	Authorised representative
IVD	In-vitro diagnostic medical device	ⵀ	For performance evaluation
∇	Number of tests		

Figure 10.9 Symbols used on medical devices.

the exception of some devices for professional use where Member States may accept English. The use of symbols is encouraged to overcome the difficulties of trying to fit multiple languages on a device label. Examples of harmonised symbols are illustrated in Figure 10.9. The labelling must also bear the CE mark of conformity (see Figure 9.2). Where applicable, this must be accompanied by the identification number of the Notified Body responsible for performing the tasks required under the different conformity assessment procedures.

10.4.10
Competent Authority Notifications and the European Databank

The manufacturer, or his/her authorised representative, must inform the Competent Authority of:

- the Member State in which he/she has his/her registered place of business;
- of his/her address; and
- a description of the devices being placed on the market (the label and instructions for use may be used as a basis).

In the case of IVDs that require the involvement of a Notified Body, data on performance characteristics, the results of performance evaluation and device certificates issued by the Notified Body must be included in the submission. The Competent Authorities are responsible for entering the following information in a European medical device databank, EUDAMED:

- Data relating to the registration of manufacturers and devices
- Data relating to device certifications issued by Notified Bodies
- Data related to vigilance procedures
- Data related to clinical investigations

The manufacturer must also notify the Competent Authority of any significant modifications to devices, or discontinuance in the market.

10.5
Placing of Devices on the Market in the US

As mentioned in Chapter 9, device classification is central to determining the regulatory path to market for a device in the US. Most Class I (and also a few Class II) devices are exempt from any device-specific registration requirements. To verify this, the manufacture needs only to consult the database of existing devices, to identify an equivalent device, and check that it is exempt from a pre-market notification requirement. If this is confirmed, then all that is required is that the manufacturer registers his/her establishment with the FDA, complies with applicable Quality System Regulation (QSR) good manufacturing practices, labels the devices in accordance with the regulations, and undertakes to report adverse events to the FDA. All other devices will have to go through a pre-market approval (PMA) procedure in the case of Class III devices, or a less-arduous 510(k) pre-market notification process in all other cases (i.e. Class I and Class II devices that are subject to pre-market notification requirements).

10.5.1
510(k) Pre-market Notification

The 510(k) pre-market notification process is not as onerous as the pre-market approval procedure, as clearance to market a device is not based on actual assessment of the safety and effectiveness of the particular device in question. Instead, devices can be cleared on the basis that they are substantially equivalent to existing devices that have been recognised as safe and effective, or that they conform to specific device standards promulgated or recognised by the FDA. There are four procedural variations to the 510(k) notification process.

10.5.1.1   Traditional 510(k)

Traditional 510(k) notifications are based on making a declaration of substantial equivalence to an existing device. This can be either a device that was on the market before the introduction of the medical device regulations in 1976, and came under the scrutiny of classification panels at that stage; or devices that were subsequently cleared using the 510(k) procedure. The identification of a suitable reference or "predicate" device can be a crucial step when embarking on this approach. The two pillars of substantial equivalence are "intended use" and "technological characteristics". However, the two devices do not have to be identical, and a degree of flexibility is tolerated – particularly with regards to methodology – where technical advances could mean that the new device is superior to the predicate device in terms of reliability, safety or effectiveness. If there is doubt as to the suitability of a predicate device the FDA Office of Device Evaluation may be consulted. The manufacturer must then collect data that demonstrates the equivalence of the new device to the predicate device. This will usually require some level of direct comparison of performance of the two devices, whether it be a laboratory, field trial or clinical investigation.

10.5.1.2 Abbreviated 510(k)

An abbreviated 510(k) is based on making a declaration of conformity to a recognised standard, special control, or specific FDA guidance. Again, the manufacturer must provide test data in support of this assertion.

10.5.1.3 Special 510(k)

If a manufacturer has modified their own device they can avail of a special 510(k) procedure for declaring substantial equivalence to their existing device, provided that the intended use or the basic technology has not changed. They must apply design controls and risk analysis to the development process, but the advantage is that they can receive a faster review process.

10.5.1.4 De Novo 510(k)

If a manufacturer has developed a new device for which no suitable predicate device exists, he/she may be able to use a *de novo* 510(k) notification procedure. This is not a common approach, as new innovative devices will tend to be viewed as Class III devices requiring pre-market approval. However, if the manufacturer can show that the level of risk does not warrant placing it in the higher risk category, he/she may petition the FDA to have it reviewed through the *de novo* 510(k) procedure.

10.5.2
Notification and Review Procedures

510(k) notifications must be submitted to the Office of Device Evaluation within the FDA CDRH. The contents of a typical 510(k) submission are outlined in Figure 10.10. The FDA are allowed 90 days to review a Traditional or Abbreviated 510(k) notification, and just 30 days for a Special 510(k). With the introduction of the Medical Device User Fee and Modernization Act of 2002, provision was made for the participation of third-party organisations in the review process. This represents a partial adoption of the concept of Notified Bodies, which prevails in Europe. The FDA have accredited a number of commercial organisations to conduct primary 510(k) reviews of 670 types of device. The FDA must then give a final determination within 30 days of receipt of the recommendation of a third-party reviewer. Because they are commercial, third-party reviewers will seek to offer faster review times in return for their review fee. If using a third-party reviewer, the FDA user fee does not apply. The outcome of a successful 510(k) notification is a letter from the FDA clearing the device for commercial sale.

10.5.3
Pre-market Approval (PMA)

The pre-market approval process is a considerably more complex procedure, as the FDA can only grant marketing approval on the basis of an assessment of the actual safety and effectiveness of the device in question. Thus, it is similar to a drug

Application Form
A 4-page form that collects basic information on the notification

Covering Letter
Identifies the device and gives a brief outline of the device

Table of Contents

User Fee Information
A copy of a completed Medical Device User Fee Coversheet, which permits checking that the review fees have been paid

Statement of Substantial Equivalence / Conformity
Identifies reference device or standard and rationale for claiming equivalence / conformity.

Labelling
Copies of labels, instructions for use or user manuals or other relevant information

Comparative information
Data demonstrating equivalence and performance. May also include information from predicate device to show equivalence of claims, etc.

Biocompatibility Assessment (if applicable)
Data establishing the biocompatibility of the materials

Truthful and Accuracy Statement
A signed declaration by a person responsible for the submission as to the truthfulness and accuracy

Shelf Life (if applicable)
Data establishing the stability of the device, accelerated stress data is acceptable

Indication for use form
Formal clarification of the indications for use, which will be made available to the public

510(k) Summary
A summary of the submission, which will be made available to the public

Figure 10.10 Outline of the content of a 510 K submission.

application, in that the submission must contain a detailed manufacturing section describing the methods for production and testing of the device, together with extensive data to substantiate its safety and effectiveness. Whereas a 510(k) application may be cleared on the basis of limited or no clinical data, a PMA requires substantial clinical investigation data. The main requirements of a PMA submission are outlined in Figure 10.11. It is advised to consult with the FDA in advance of finalising a PMA strategy if there is any uncertainty about required data, particularly in the case of significantly innovative devices. The FDA may refer to an advisory council in such circumstances. Once a submission is received, it will go through a screening procedure to check that it is complete for filing. A filing decision must be made within 45 day. The FDA are allowed 180 day to review and come to a decision on the submission. This usually results in the issue an approvable letter, which will just identify remaining minor issues that need to be resolved for the FDA to issue an

1. Name & Address of Applicant

2. Table of Contents

3. Summary
Including indications for use, device description, alternative practices and procedures currently available for the condition, marketing history (US or foreign) non-clinical laboratory studies, clinical investigations and overall conclusions.

4. Device description and Manufacture
Detailed drawings and description of device and methods used for manufacture and control of the device.

5. Reference to Standards
Reference to standards for radiation protection and any other appropriate voluntary standards relating to safety or effectiveness.

6. Technical Sections
Non-clinical laboratory studies used to investigate microbiological, toxicological, immunological, biocompatibility, stress, wear, shelf life, and other characteristics of the device.
Data from clinical investigations on human subjects, plus statements as regards investigators, IRBs and informed consent.

7. Single Clinical Investigation Justification
Information to justify reliance on a single clinical investigation, where applicable.

8. Bibliography
Relevant published reports and commentary.

9. Device Samples (on request)
Provide samples of the device or a location where they may be tested by the FDA.

10. Copies of Labelling
Labels, instructions for use, user manuals, etc.

11. Environmental Assessment
Assessment or declaration of exclusion from requirement.

12. Financial Disclosure
Declarations of investigator's financial interests, if any.

Figure 10.11 Required content of a PMA application.

approval order, or an order denying approval, if the FDA finds major deficiencies in the submission. As part of an approval order the FDA may impose additional requirements such as batch testing, or additional labelling requirements.

10.5.4
Changes to a PMA-Approved Device

Changes to an approved device that could affect its safety or effectiveness must be submitted to the FDA as a supplemental application. These include changes to the indications for use, changes to the labelling, packaging or sterilisation procedures, new manufacturing or packaging sites, changes to the device or its specifications and alterations to its shelf life, if not based on a protocol agreed with the FDA. The procedure for submission and review of a supplement are the same as for the initial PMA, but in practice are likely to be faster because of their limited content. Minor changes that enhance device safety or just make editorial corrections to labelling may be introduced without prior approval, provided that the changes are notified to the FDA as part of periodic reports.

10.5.5
Humanitarian Use Devices

Similar to Orphan Drug provisions, a device may be designated as a Humanitarian Use Device (HUD) if it is intended to diagnose or treat a rare disease or condition affecting less than 4000 US sufferers. Applications for HUD status must be submitted to the Office for Orphan Products Development, together with information supporting the applications. A decision as to status will be made within 45 days. The advantage of HUD status is that permission to market the device may be granted on the basis of a Humanitarian Device Exemption (HDE) application, which does not require the sponsor to submit comprehensive data as to the effectiveness of the device. The permitted review times are shorter at 30 days for a filing decision and 75 days for an evaluation decision.

10.5.6
Labelling of Devices

General labelling requirements for devices are set out in 21 CFR 801. These include the name and place of business of the manufacturer, packer or distributor, text sizes, sufficient instructions to enable the device to be used correctly, and information on quantity per package, where applicable. This part also contains labelling requirements for particular types of device. For example, devices available only on prescription should bear the statement, "Caution: Federal law restricts this device to sale by or on the order of a—", the blank to be filled with the word "physician", "dentist", "veterinarian" or other licensed professional. Specific requirements for the labelling of *in-vitro* diagnostic reagents may be found in 21 CFR 809. Devices approved under an HDE must bear the statement, "Humanitarian Device. Authorized by Federal law for use in the [treatment or diagnosis] of [specify disease or condition]. The effectiveness of this device for this use has not been demonstrated".

10.6
Chapter Review

Although there is significant difference in the detail, the regulatory requirements for evaluation and marketing authorisation of devices in Europe and the US both reflect a tiered response in relation to risk. This is illustrated in Table 10.4. In Europe

Table 10.4 Overview of requirements versus risk according to EU and US regulations.

	EU Requirements Overview			
Risk	**Clinical Data**	**Authorisation of Clinical Investigations**	**Instructions for Use**	**Conformity Assessment Procedure**
High (III)	Specific Clinical Investigations expected	60 days for approval by CA	Required	Notified Body certification of specific device (Type or design)
Moderate (II)	Review of existing clinical data may be acceptable	60 days for approval by CA for certain devices	Not required, if can be used safely (except IVDs)	Notified Body involvement
Low (I)	Usually possible to justify exclusion of clinical data	Proceed unless CA raise objections	Not required, if can be used safely, (except IVDs)	Self assessment by manufacturer

	US Requirements Overview		
Risk Class	**Device Evaluation**	**Quality System Regulations**	**Marketing Authorisation**
III	IDE	Full QSR	PMA
II	Abbreviated IDE possible	Full QSR	510(k)
I	May be IDE exempt	QSR without design control	Generally 510(k) exempt

conformity assessment is the responsibility of Notified Bodies, except for low-risk devices where the manufacturer himself/herself can perform the assessment. The Competent Authorities are only likely to become involved should problems be encountered in the market place. In the US, although third-party review is now possible for lower-risk devices, the ultimate responsibility for market clearance or approval rests with the FDA.

10.7
Further Reading

- European Directives
  Directive 90/385/EEC
  Directive 93/42/EEC
  Directive 98/79/EC
  http://ec.europa.eu/enterprise/medical_devices/legislation_en.htm.
- MEDDEV Guides
  2.7 - Clinical data
  2.14/3 - IVD Instructions for use
  http://ec.europa.eu/enterprise/medical_devices/meddev/index.htm.
- US Code of Federal Regulations
  Labelling 21 CFR Parts 801 & 809
  Pre market Notification – 21 CFR Part 807, subpart E
  Investigational Device Exemptions - 21 CFR Part 812   Pre market approval – 21 CFR Part 814
  Quality System Regulations - 21 CFR Part 820
  www.fda.gov.

11
Good Manufacturing Practice (GMP)

11.1
Chapter Introduction

The previous chapters examined the considerable efforts that go into establishing that medical products as developed are safe and effective. However, this alone will not assure the safety of public health. It is equally important to address the manufacturing operations to ensure that the manufactured product displays a high level of purity and consistency. Fundamental to this, are the principles of Good Manufacturing Practice (GMP), which have been developed by regulatory authorities. These are based on the essential principles of quality assurance (QA) systems in general, but with specific focus given to manufacturing operations and particular requirements that may be required in order to assure the quality of drugs and devices. Thus, while the requirements of general quality systems are expressed in very generic terms, GMP principles tend to be stated as more detailed and specific requirements that reflect the particular product or operation to which they apply.

11.2
Drug GMP Regulations and Guidance

Drug GMPs have been developed as a combination of regulations and guidance. Specific regulations have been issued to define GMP requirements for the manufacture of finished drug products, whereas guidance is used to establish requirements for the earlier stage of active pharmaceutical ingredient (API) manufacture. In general, the level of control is expected to increase as manufacturing operations progress to the final stages in the process. This is particularly evident in the case for API manufacture, as illustrated in Figure 11.1. A similar approach is also witnessed in terms of regulatory control. In Europe, most regulatory vigilance is targeted at the manufacture and release of drug products, as this provides the regulators with the most effective means of controlling the quality of the product that is released to

Medical Product Regulatory Affairs. John J. Tobin and Gary Walsh
Copyright © 2008 WILEY-VCH Verlag GmbH & Co. KGaA, Weinheim
ISBN: 978-3-527-31877-3

Type of manufacturing	Application of ICH GMP guide to indicated steps (shown in grey) of API manufacturing				
Chemical manufacturing	Production of the API Starting Material	Introduction of the API Starting Material into process	Production of Intermediate(s)	Isolation and purification	Physical processing, and packaging
API derived from animal sources	Collection of organ, fluid, or tissue	Cutting, mixing, and/or initial processing	Introduction of the API Starting Material into process	Isolation and purification	Physical processing, and packaging
API extracted from plant sources	Collection of plants	Cutting and initial extraction(s)	Introduction of the API Starting Material into process	Isolation and purification	Physical processing, and packaging
Herbal extracts used as API	Collection of plants	Cutting and initial extraction		Further extraction	Physical processing, and packaging
API consisting of comminuted or powdered herbs	Collection of plants and/or cultivation and harvesting	Cutting/comminuting			Physical processing, and packaging
Biotechnology: fermentation/cell culture	Establishment of master cell bank and working cell bank	Maintenance of working cell bank	Cell culture and/or fermentation	Isolation and purification	Physical processing, and packaging
"Classical" fermentation to produce an API	Establishment of cell bank	Maintenance of the cell bank	Introduction of the cells into fermentation	Isolation and purification	Physical processing, and packaging

Increasing GMP requirements

Figure 11.1 Applicability of GMP requirements to API manufacture (from ICH Q7 guideline).

the market. However, should regulators identify problems that may be attributed to an API ingredient, a targeted inspection of the API manufacturer is likely to ensue. This strategy is illustrated in Figure 11.2. Under the US FDC Act, manufacturers of APIs come under the same regulatory controls as manufacturers of finished pharmaceutical products as the definition of a drug includes components of drug products.

The European regulations relating to drug GMPs are set out in two directives; Directive 2003/94/EC, which lays down the principles and guidelines of good manufacturing practice in respect of medicinal products for human use and investigational medicinal products for human use.

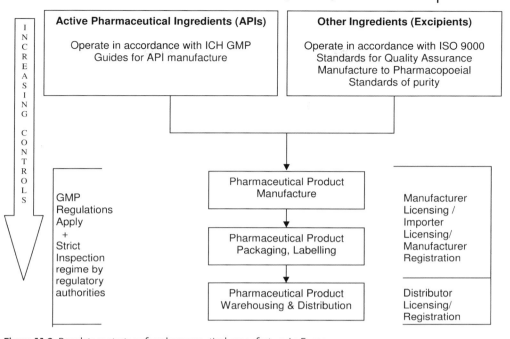

Figure 11.2 Regulatory strategy for pharmaceutical manufacture in Europe.

Directive 91/412/EEC, which lays down the principles and guidelines of good manufacturing practice for veterinary medicinal products.

The core US regulations are contained in:

Chapter 21 of the Code of Federal Regulations (21 CFR)

Part 210, current Good Manufacturing Practice in Manufacturing, Processing, Packing, or Holding of Drugs, General Requirements

Part 211, current Good Manufacturing Practice for Finished Pharmaceuticals.

Supplementary requirements for specific types of drug products are contained in other Parts of Chapters 9 and 21 of the Code of Federal Regulations, as summarised in Table 11.1. ICH Q7 guideline provides globally harmonised principles of GMP for APIs.

The topics addressed by European GMP regulations and guidance are shown in Figure 11.3. The core requirements are set down in the Articles of the directives, whereas detailed guidance on specific aspects is appended as Annexes. The Annexes, together with guidance on the core GMP requirements, may be found in The Rules Governing Medicinal Products in the European Union, Volume 4, Medicinal Products for Human and Veterinary Use: Good Manufacturing Practice. The topics covered by the basic US regulations for drug GMPs are shown in Figure 11.4, while the headings of the ICH guideline for API GMPs are shown in Figure 11.5. Although the terminology and sequence of topics may appear different, the essential principles are the same in all cases. As will become apparent, many of these measures are

Table 11.1 US current Good Manufacturing Practice (cGMP) regulations for specific types of drug products.

CFR No.	Subject Matter
9 CFR	**Veterinary biologics**
Part 108	Facility requirements for licensed establishments
Part 109	Sterilization and pasteurization at licensed establishments
Part 113	Standard requirements
Part 114	Production requirements for biological products
21 CFR	**Drugs and human biologics**
Part 225	current Good Manufacturing Practice for Medicated Feeds
Part 226	current Good Manufacturing Practice for Type A Medicated Articles
Part 600	Biologic Products General
Part 606	current Good Manufacturing Practice for Blood and Blood Components
Part 610	General Biological Product Standards
Part 630	General Requirements for Blood, Blood Components and Blood Products
Part 640	Additional Standards for Human Blood and Blood Products
Part 660	Additional Standards for Diagnostic Substances for Laboratory Tests
Part 680	Additional Standards for Miscellaneous Products

directed at ensuring that the purity and authenticity of drug products is guaranteed and represent specific embodiments of the general principles of quality assurance systems that were discussed earlier in Chapter 2.

11.3
Essential GMP Requirements

11.3.1
Quality Assurance System

At its core, each manufacturer must have a QA system appropriate to the manufacture of pharmaceutical products. Essential elements that such a system should address are summarised in Figure 11.6.

11.3.2
Personnel

Previously, Chapter 2 examined the importance of the organisation, training and educational qualifications of personal as core elements of quality systems in general. *Personal hygiene* is a specific additional aspect that must be addressed to prevent the contamination of drugs. Training and practices must be adopted so as to achieve:

- high standards of personal hygiene;
- the correct use of appropriate clothing such as gowns, hairnets or masks;
- a prohibition of eating, drinking, chewing, smoking or storage of food drinks or personal items in manufacturing areas;

Core Regulatory Requirements (Articles)
1. **Quality assurance system** – maintain an effective pharmaceutical QA system
2. **Personnel** – organisation, responsibilities, training, qualifications, hygiene
3. **Premises and equipment** – suitability, layout, maintenance, qualification and validation
4. **Documentation** – procedures, specifications, retention of records, electronic data
5. **Production** – follow procedures, record deviations, prevent mix-up, validate processes
6. **Quality control** – responsibility, review test results and production records, retain samples
7. **Work contracted out** – contract clearly defining responsibilities, subject to GMP inspections
8. **Complaints, product recall and emergency unblinding** – complaints and recall procedures
9. **Self-inspection** – regular internal auditing and follow up corrective actions

Guidance Annexes
Annex 1 Manufacture of sterile medicinal products
Annex 2 Manufacture of biological medicinal products for human use
Annex 3 Manufacture of Radiopharmaceuticals
Annex 4 Manufacture of Veterinary Medicinal Products other than Immunologicals
Annex 5 Manufacture of Immunological Veterinary Medicinal Products
Annex 6 Manufacture of Medicinal Gases
Annex 7 Manufacture of Herbal Medicinal Products
Annex 8 Sampling of Starting and Packaging Materials
Annex 9 Manufacture of Liquids, Creams and Ointments
Annex 10 Manufacture of Pressurised Metered Dose Aerosol Preparations for Inhalation
Annex 11 Computerised Systems
Annex 12 Use of Ionising Radiation in the Manufacture of Medicinal Products
Annex 13 Manufacture of Investigational Medicinal Products
Annex 14 Manufacture of Products derived from Human Blood or Human Plasma
Annex 15 Qualification and validation
Annex 16 Certification by a Qualified person and Batch Release
Annex 17 Parametric Release
Annex 18 Good manufacturing practice for active pharmaceutical ingredients (The ICH Guide)
Annex 19 Reference and Retention Samples

Figure 11.3 Basic topics addressed by the EU GMP Directives for medicinal products.

- restrictions on personnel with infectious diseases or other relevant conditions from working on the products; and
- the avoidance of direct contact with the product.

11.3.3
Premises and Equipment

The careful planning of the layout and construction of premises is another GMP requirement that is key to reducing the risk of contamination or product mix-up. Ideally, the layout should be designed so as to allow the production to take place in areas connected in a logical order, corresponding to the sequence of the operations and the requisite levels of cleanliness. The design should minimise the movement of personnel or equipment through the production areas. Areas where raw materials are weighed or charged into reactors should be physically separated from other production areas so as to reduce potential cross-contamination by dusts generated during such operations. The manufacture of certain critical products such as penicillins or

Subpart A--General Provisions
Sec. 211.1 Scope.
Sec. 211.3 Definitions.

Subpart B--Organization and Personnel
Sec. 211.22 Responsibilities of quality control unit.
Sec. 211.25 Personnel qualifications.
Sec. 211.28 Personnel responsibilities.
Sec. 211.34 Consultants.

Subpart C--Buildings and Facilities
Sec. 211.42 Design and construction features.
Sec. 211.44 Lighting.
Sec. 211.46 Ventilation, air filtration, air heating and cooling.
Sec. 211.48 Plumbing.
Sec. 211.50 Sewage and refuse.
Sec. 211.52 Washing and toilet facilities.
Sec. 211.56 Sanitation.
Sec. 211.58 Maintenance.

Subpart D—Equipment
Sec. 211.63 Equipment design, size and location.
Sec. 211.65 Equipment construction.
Sec. 211.67 Equipment cleaning and maintenance.
Sec. 211.68 Automatic, mechanical and electronic equipment.
Sec. 211.72 Filters.

Subpart E--Control of Components and Drug Product Containers and Closures
Sec. 211.80 General requirements.
Sec. 211.82 Receipt and storage of untested components, drug product containers and closures
Sec. 211.84 Testing and approval or rejection of components, drug product containers and closures.
Sec. 211.86 Use of approved components, drug product containers and closures.
Sec. 211.87 Retesting of approved components, drug product containers and closures.
Sec. 211.89 Rejected components, drug product containers and closures.
Sec. 211.94 Drug product containers and closures.

Subpart F--Production and Process Controls
Sec. 211.100 Written procedures; deviations.
Sec. 211.101 Charge-in of components.
Sec. 211.103 Calculation of yield.
Sec. 211.105 Equipment identification.
Sec. 211.110 Sampling and testing of in-process materials and drug products.
Sec. 211.111 Time limitations on production.
Sec. 211.113 Control of microbiological contamination.
Sec. 211.115 Reprocessing.

Subpart G--Packaging and Labelling Control
Sec. 211.122 Materials examination and usage criteria.
Sec. 211.125 Labelling issuance.
Sec. 211.130 Packaging and labelling operations.
Sec. 211.132 Tamper-evident packaging requirements for over-the-counter (OTC) human drug products.
Sec. 211.134 Drug product inspection.
Sec. 211.137 Expiration dating.

Subpart H--Holding and Distribution
Sec. 211.142 Warehousing procedures.
Sec. 211.150 Distribution procedures.

Figure 11.4 Core GMP regulations as set out in Chapter 21 of the US Code of Federal Regulations, Part 211 (21 CFR 211).

Subpart I--Laboratory Controls
Sec. 211.160 General requirements.
Sec. 211.165 Testing and release for distribution.
Sec. 211.166 Stability testing.
Sec. 211.167 Special testing requirements.
Sec. 211.170 Reserve samples.
Sec. 211.173 Laboratory animals.
Sec. 211.176 Penicillin contamination.

Subpart J--Records and Reports
Sec. 211.180 General requirements.
Sec. 211.182 Equipment cleaning and use log.
Sec. 211.184 Component, drug product container, closure, and labeling records.
Sec. 211.186 Master production and control records.
Sec. 211.188 Batch production and control records.
Sec. 211.192 Production record review.
Sec. 211.194 Laboratory records.
Sec. 211.196 Distribution records.
Sec. 211.198 Complaint files.

Subpart K--Returned and Salvaged Drug Products
Sec. 211.204 Returned drug products.
Sec. 211.208 Drug product salvaging.

Figure 11.4 (*Continued*)

highly cytotoxic drugs may require dedicated facilities to eliminate all risks of cross-contamination. Consideration must be given to providing heating, ventilation and air conditioning (HVAC) systems that are in keeping with the operations undertaken in a particular area. Such measures should be complemented by surfaces on floors, walls and ceilings that facilitate cleaning – hard smooth surfaces with minimal recesses. In general, such measures may be appropriate for the final stages in a manufacturing process (such as packaging), where the highest levels of hygiene are expected. A pest control programme must be applied throughout the facility. There must be adequate storage space to allow clear segregation between raw materials, intermediates or finished products that are approved, and those that are on hold pending Quality Control (QC) approval, or that have been rejected. Quality Control laboratories, personnel lockers, toilets, and canteen facilities should be kept separate from the production areas. A schematic diagram of a facility layout is shown in Figure 11.7.

The manufacture of sterile products requires special measures in terms of facilities and arrangements to ensure that the products can be handled with minimal risk of microbial contamination. The generation of airflows of clean air that has been filtered, usually through high-efficiency particulate air (HEPA) filters, is fundamental to this objective. This is combined with surfaces that are specifically designed for easy cleaning, and strict operational practices that minimise the risk of contamination. Different levels of cleanliness will need to be achieved depending on the type of operation carried out. Four grades are generally recognised in pharmaceutical manufacture. These are based on the levels of dust particles and microbes tolerated in an area. The specifications according to European guidelines are summarised in Table 11.2. Areas must be monitored under two conditions: (i) the at-rest condition, when the air circulation system is switched on but activities are not in progress; and

1 Introduction
1.1 Objective
1.2 Regulatory Applicability
1.3 Scope

2 Quality Management
2.1 Principles
2.2 Responsibilities of the Quality Unit(s)
2.3 Responsibility for Production Activities
2.4 Internal Audits (Self-Inspection)
2.5 Product Quality Review

3 Personnel
3.1 Personnel Qualifications
3.2 Personnel Hygiene
3.3 Consultants

4 Buildings and Facilities
4.1 Design and Construction
4.2 Utilities
4.3 Water
4.4 Containment
4.5 Lighting
4.6 Sewage and Refuse
4.7 Sanitation and Maintenance

5 Process Equipment
5.1 Design and Construction
5.2 Equipment Maintenance and Cleaning
5.3 Calibration
5.4 Computerised Systems

6 Documentation and Records
6.1 Documentation System and Specifications
6.2 Equipment Cleaning and Use Record
6.3 Records of Raw Materials, Intermediates, API Labelling and Packaging Materials
6.4 Master Production Instructions (Master Production and Control Records)
6.5 Batch Production Records (Batch Production and Control Records)
6.6 Laboratory Control Records
6.7 Batch Production Record Review

7 Materials Management
7.1 General Controls
7.2 Receipt and Quarantine
7.3 Sampling and Testing of Incoming Production Materials
7.4 Storage
7.5 Re-evaluation

8 Production and In-Process Controls
8.1 Production Operations
8.2 Time Limits
8.3 In-process Sampling and Controls
8.4 Blending Batches of Intermediates or APIs
8.5 Contamination Control

9 Packaging and Identification Labelling of APIs and Intermediates
9.1 General
9.2 Packaging Materials

Figure 11.5 Headings of ICH GMP guideline for Active Pharmaceutical Ingredients (API) manufacture.

9.3 Label Issuance and Control
9.4 Packaging and Labelling Operations

10 Storage and Distribution
10.1 Warehousing Procedures
10.2 Distribution Procedures

11 Laboratory Controls
11.1 General Controls
11.2 Testing of Intermediates and APIs
11.3 Validation of Analytical Procedures
11.4 Certificates of Analysis
11.5 Stability Monitoring of APIs
11.6 Expiry and Retest Dating
11.7 Reserve/Retention Samples

12 Validation
12.1 Validation Policy
12.2 Validation Documentation
12.3 Qualification
12.4 Approaches to Process Validation
12.5 Process Validation Program
12.6 Periodic Review of Validated Systems
12.7 Cleaning Validation
12.8 Validation of Analytical Methods

13 Change Control

14 Rejection and Reuse of Materials
14.1 Rejection
14.2 Reprocessing
14.3 Reworking
14.4 Recovery of Materials and Solvents
14.5 Returns

15 Complaints and Recalls

16 Contract Manufacturers (including Laboratories)

17 Agents, Brokers, Traders, Distributors, Repackers, and Relabellers
17.1 Applicability
17.2 Traceability of Distributed APIs and Intermediates
17.3 Quality Management
17.4 Repackaging, Relabelling and Holding of APIs and Intermediates
17.5 Stability
17.6 Transfer of Information
17.7 Handling of Complaints and Recalls
17.8 Handling of Returns

18 Specific Guidance for APIs Manufactured by Cell Culture/Fermentation
18.1 General
18.2 Cell Bank Maintenance and Record keeping
18.3 Cell Culture/Fermentation
18.4 Harvesting, Isolation and Purification
18.5 Viral Removal/Inactivation Steps

19 APIs for Use in Clinical Trials
19.1 General
19.2 Quality
19.3 Equipment and Facilities

Figure 11.5 (*Continued*)

19.4 Control of Raw Materials
19.5 Production
19.6 Validation
19.7 Changes
19.8 Laboratory Controls
19.9 Documentation

20 Glossary

Figure 11.5 (*Continued*)

(ii) the in-operation condition, where activities are in progress with the full complement of operators.

Grade A is the highest standard, and is required for local zones where high-risk tasks are carried out. Examples are the aseptic filling of a product following filtration through a 0.22 µm filter to render it sterile, and general aseptic manipulations. This is usually achieved by the use of cabinets or hoods that enable laminar airflow patterns to be established.

The general environment around such stations should also operate to recognised standards of cleanliness. This involves the creation of enclosed clean room areas in which filtered air is circulated at a positive pressure relative to atmospheric pressure. Personnel and equipment entry is through air-lock chambers. The generated pressure differentials ensure that there will be an outflow of clean air each time a door or hatch is opened, thus reducing the risk of contaminants entering the area. The chambers also serve as gowning areas for changing into specific protective

1. Pharmaceutical products are designed and developed in a way that takes account of the requirements of Good Manufacturing Practice and Good Laboratory Practice.

2. Production and control operations are clearly specified and Good Manufacturing Practice adopted.

3. Managerial responsibilities are clearly specified.

4. Arrangements are made for the manufacture, supply and use of the correct starting and packaging materials.

5. All necessary controls on intermediate products, and any other in-process controls and validations are carried out.

6. The finished product is correctly processed and checked, according to the defined procedures.

7. Pharmaceutical products are not sold or supplied before a Qualified Person has certified that each production batch has been produced and controlled in accordance with the requirements of the Marketing Authorisation and any other regulations relevant to the production, control and release of medicinal products.

8. Satisfactory arrangements exist to ensure, as far as possible, that the pharmaceutical products are stored, distributed and subsequently handled so that quality is maintained throughout their shelf life.

9. There is a procedure for Self-Inspection and/or quality audit, which regularly appraises the effectiveness and applicability of the Quality Assurance system.

Figure 11.6 Pharmaceutical quality assurance system principles.

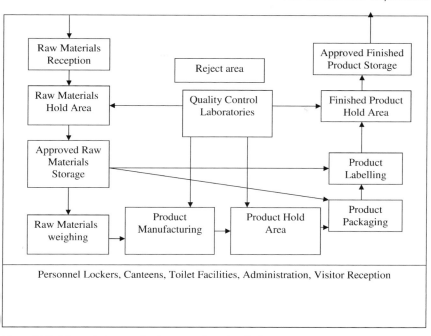

Figure 11.7 Schematic layout of a drug-manufacturing facility.

clothing. The type of clothing will vary depending on the level of cleanliness to which an area operates – at the top end fresh sanitised clothing is used for each work session. The entry of personnel and equipment should be kept to the minimum necessary to carry out the operation. An idealised clean room layout is shown in Figure 11.8.

Clean rooms can be designed to operate to *Grade B*, *Grade C* or *Grade D* levels of cleanliness, depending on the nature of the activity undertaken. For example, Grade B is the recommended background environment for aseptic preparation and filling, a Grade C environment is suitable for preparation of solutions to be filtered or for filling of products that will be sterilised after filling, and a Grade D environment is suitable for general support activities such as handling of components after washing.

Table 11.2 Specifications for air quality according to European guidelines.

| | | At-rest | | In-operation | |
| | | Maximum permitted number of particles m^{-3} | | | |
Grade	cfu m^{-3}	0.5 μm	5 μm	0.5 μm	5 μm
A	<1	3500	1	3500	1
B	<10	3500	1	350 000	2000
C	<100	350 000	2000	3 500 000	20 000
D	<200	3 500 000	20 000	Depends on activity	Depends on activity

cfu = colony forming units.

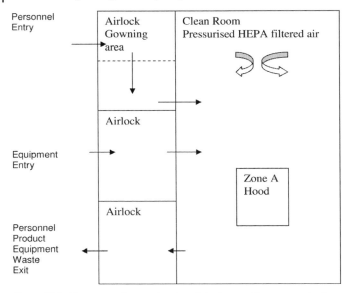

Figure 11.8 Clean room layout showing air-lock areas for entry and exit.

The ability to achieve different clean room grades is determined largely by the number of air changes relative to the size of the room and the number of personnel and equipment in the room, and the efficiency of the air filters.

Operations involving infectious agents or organisms will require further measures to ensure bio-security and safety. This usually involves the use of biosafety cabinets, which are designed to protect both the product and the user from contamination. Appropriate disinfection and bio-containment procedures must be adopted to prevent unwanted release of dangerous organisms.

An essential utility for most sites is a *water purification system* that can generate water of appropriate quality. Two grades of water are used in pharmaceutical production:

- Purified Water, which may be used for cleaning/rinsing or to prepare oral-dosage drugs
- Water-for-Injection, which is used for parenteral injection products.

The specifications according to the US Pharmacopoeia are shown in Table 11.3. Purified Water may be produced by a combination of filtration, ion exchange, reverse osmosis and ozone sterilisation, whereas Water for Injection requires an additional distillation step. The water is then distributed through closed-loop pipework, where it is kept constantly circulating to prevent stagnation. The pipework or tubing from take-off valves should be of minimum length (less than six times the diameter) and be devoid of kinks or loops that prevent water from draining away under gravity. The system must be regularly monitored to ensure that the required standards of purity and sterility are maintained.

Table 11.3 Specifications for United States Pharmacopoeia (USP) grade Purified Water and Water for Injection.

Parameter	Purified water	Water for Injection
pH	5.0–7.0	5.0–7.0
Conductivity ($\mu S\,cm^{-1}$)	$\leq$1.1 at 20 °C	$\leq$1.1 at 20 °C
Total organic carbon (parts per billion)	<500	<500
Microbial	<100 cfu mL^{-1}	<10 cfu mL^{-1}
Endotoxin (EU mL^{-1})	–	0.25

cfu = colony forming units.

11.3.4
Documentation

Documentation is at the core of all quality systems, and was discussed in Chapter 2. Examples of the types of documentation associated with pharmaceutical production are shown in Figure 11.9. European regulations require that records which permit tracing of the full history for each batch of product should be retained for 1 year after expiry of the product or 5 years, whichever is the longer. US regulations require the retention of records for 1 year after batch expiry or 3 years after the last distribution in cases of some OTC products where no expiry date is assigned. Additionally, the US regulations require the preparation of a "Master Production and Control Record",

Specifications for:
- starting materials
- packaging materials
- intermediates
- bulk product
- finished product
Manufacturing formulae
Processing instructions
Packaging instructions
Procedures for:
- receipt, sampling and testing of materials
- validation
- equipment assembly and calibration
- maintenance, cleaning and sanitation
- personnel matters including training, clothing, hygiene
- environmental monitoring
- pest control
- complaints
- recalls
- returns
Records for:
- material receipt
- batch processing
- batch packaging
- testing
- distribution
and the other activities listed above

Figure 11.9 Examples of documentation associated with pharmaceutical manufacture.

which identifies all the instructions necessary to produce a batch (specifications, batch formulae, manufacture and test procedures, packaging, labelling, etc.).

11.3.5
Production

Manufacturing operations must be carried out in such a way as to minimise the possibility of cross-contamination or mix up. All materials must be clearly identified at all stages in the process. Particular attention is given to packaging and labelling operations where a history of past mishaps has underscored the potential fatal consequences of product mix up. Where different products are processed on the same line, there should be strict line clearance of the first product before commencing to process the next product. The issuing of labels must be strictly controlled and recorded. Label reconciliations should be performed at the end of an operation to ensure that the number of labels used match the number of products that should have been labelled. Unused labels should be either returned to the approved storage area, or destroyed, if printed with batch-specific information. At all appropriate manufacturing stages actual yields should be calculated and compared to anticipated yield. Deviations outside stated tolerances should be investigated before proceeding. Necessary deviations should be authorised using established deviation procedures that will usually involve the quality department. It is standard practice for all critical steps to be double-checked by a second operator.

11.3.6
Quality Control

A quality control group must be established independent of all other areas, particularly production. The release of product must be based on a combination of the review of test results and the batch manufacturing, packaging and labelling records. Sampling of materials for testing (raw materials, intermediates or finished product) should be based on statistically valid sampling plans that are consistent with the previous history of the material; that is, problematic materials will require significantly more sampling than materials that have never had a history of problems. Samples of finished drug product should be retained for 1 year after the expiry of the product, or 3 years after last distribution in the case of products that have no expiry date (e.g. some OTC products in the US). In Europe, samples of all starting materials other than solvents gases and water should be retained for 2 years. US regulations require the retention of active pharmaceutical ingredients for 1 year after the expiry of the last batch of finished product in which they were used.

11.3.7
Work Contracted Out

Pharmaceutical manufacture may involve the contracting out of certain operations in the process, such as packaging and labelling, terminal sterilisation of products or performing specific analytical procedures. All such activities must be the subject of contracts that clearly define the responsibilities and duties of the parties involved. The contracted party must comply with GMP regulations and can be subject to inspection by the regulatory authorities. They may not subcontract any part of the work without the knowledge and prior approval of the original contract-giver.

11.3.8
Complaints, Product Recall and Emergency Un-Blinding

A system must be established to thoroughly investigate all complaints. If the situation so demands, there must be procedures in place to effectively recall a product from the market. This requires the retention of distributor contact details and precise distribution records so that the destination of each shipment can be determined. The regulatory authorities must be informed of any planned recall actions (see Chapter 12). Procedures are also required for emergency un-blinding of materials undergoing clinical trial. The responsibility for complaints and the initiation of product recalls should be assigned to designated individuals.

11.3.9
Self-Inspection

Regular internal audits of GMP operations should be conducted according to an *audit master schedule*. Records of the findings and any follow-up actions should be maintained. Audits of key suppliers should also be conducted as part of a vendor qualification programme.

Collectively, the combination of appropriate facilities, equipment, documentation, manufacturing practices and quality control procedures provide a basis for effective product and process control. This is illustrated in Figure 11.10.

11.4
Validation

Validation is key to establishing most critical GMP operations. Typical systems that require validation are as follows:

- Facilities
- Equipment
- Manufacturing process
- Computer systems
- Analytical methods
- Cleaning methods
- Sterilisation methods
- Water purification system

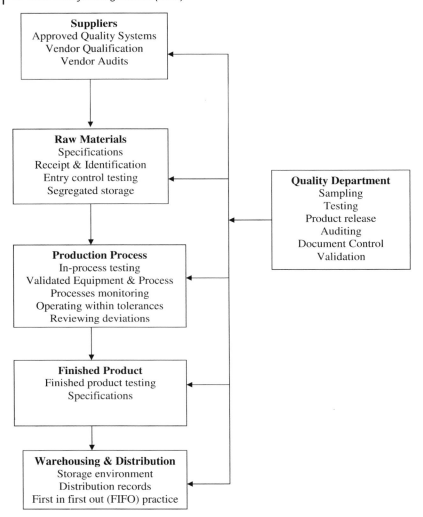

Figure 11.10 Schematic showing some of the elements that go into achieving effective product and process control.

Pharmaceutical sites will usually create a dedicated team of validation specialists to coordinate all validation activities. They should operate according to a validation master plan that has been developed using risk analysis to identify the most critical systems requiring validation/re-validation. Before validating a system or process, a written protocol should be prepared that describes the system, the critical aspects, the objectives, the test methods and the acceptance criteria that will be applied. A validation report should be prepared on completion of each protocol.

11.4.1
Facilities and Equipment Validation

The validation of new facilities, systems or equipment is usually accomplished using a four-stage process of design qualification (DQ), installation qualification (IQ), operational qualification (OQ) and performance qualification (PQ).

Design qualification is of most relevance for systems that involve custom-design such as the facilities themselves, where it is necessary to demonstrate that the design complies with GMP requirements. For simpler off-the-shelf equipment it is usually only a matter of selecting the correct equipment for the intended use.

Installation qualification involves performing checks to ensure that the correct equipment or system has been installed and/or connected, including all necessary controls, monitors, instrumentation, or ancillary services. These checks should include verification that relevant operator manuals or instructions have been received from the supplier and that any applicable calibration steps have been identified.

Operational qualification involves performing a series of tests to check that all elements of the system are functional across the specified operating range. This usually involves performing challenges at the "worst case" extreme operating conditions. The process should allow confirmation of final operation, maintenance and calibration procedures.

Finally, *performance qualification* is carried out using the materials and range of operating conditions that are envisaged under normal use. Consider, by way of example, a new filling line that uses a peristaltic pump and tubing to dispense specified amounts of liquid. The reproducibility of the dispensed volume could be influenced by the size of the tubing, the viscosity of the liquid, changes in the feed-line pressure as the height of the reservoir tank drops, or a loss of elasticity of the tubing as it ages. The performance qualification should set out to demonstrate that by the use of appropriate diameter tubing and pump speeds, a header tank to maintain a constant pressure feed and defined calibration check and tubing replacement intervals, the system can deliver the necessary process capability to achieve the specified filling tolerances. Products representative of low and high viscosity and small and large fill volumes would be selected, and checks of the filling volumes delivered over an appropriate period of operation would be statistically analysed. The principle of process capability is illustrated in Figure 11.11.

11.4.2
Process Validation

In addition to the qualification and validation of generic systems such as filling lines or cleaning methods, where it is acceptable to evaluate the system with selected products, specific process validation protocols should be executed for each product manufactured. Again, the objective is to demonstrate that the product can be manufactured to the required specifications under the normal range of prevailing operating conditions. In general, the satisfactory manufacture of three consecutive batches is considered sufficient for this purpose. Process validation should normally

PROCESS CAPABILITY

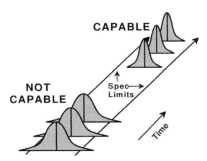

Figure 11.11 A graphical illustration of the concept of process capability (from GHTF Guidance: Quality Management Systems – Process Validation).

be completed before commercial sales commence (prospective validation). Occasionally, concurrent validation may be acceptable, but will require documented justification of why it is necessary. Retrospective validation may be applied to old stable processes that were established before the concept of validation became standard practice. This may be accomplished by carrying out a review of all data from 10 to 30 batches to establish consistency.

New systems or processes may also need to be qualified from an operational safety perspective. This is particularly relevant in the case of chemical synthesis involving exothermic reactions. Critical safety aspects are usually identified using hazard operability or HAZOP assessments and studies. For example, a HAZOP analysis of an exothermic reaction vessel would involve consideration of the consequence of failure of the motors for mixers or circulation pumps for cooling water. Thus, the qualification of such a system would involve checks and assessment to ensure that the system/process can be operated safely and that pressure relief valves or other emergency measures are adequate and functional.

11.4.3
Computer Systems Validation

All computerised systems used in manufacturing must be assessed for their potential impact on GMP activities. The extent and level of validation will depend on the functions and complexity of the system involved. The Society of Pharmaceutical Engineering has developed a guide for validation of automated systems as part of Good Automated Manufacturing Practice (GAMP). A basis for categorising systems and a sample risk analysis matrix are shown in Figure 11.12 and Table 11.4.

The FDA issued specific regulations in relation to electronic records and electronic signatures in Chapter 21 of the Code of Federal Regulations, Part 11. These apply where computers are relied on to create and maintain legally required records. Theses include batch manufacturing, testing, and release records where operators are obliged to sign that they have completed or authorised a task or any regulatory submission that involves a signed declaration. The use of electronic signatures

Category 1: Operating Systems
These are commercially available operating systems (e.g. Microsoft Windows XP). They are not subject to specific validation, but features are functionally evaluated and challenged during testing of the application

Category 2: Firmware
These can be simple logic controllers that are part of instruments. Their identity and functionality is normally confirmed as part of the IQ/OQ of the instrument.

Category 3: Standard Software Packages
These are commercially available off-the-shelf packages. They are used as is, with significant reliance placed on documentation provided by the developers to ensure that the package meets the user requirements specification (URS)

Category 4: Configurable Software Packages
These are software packages that can be configured according to users needs, e.g. enterprise resource planning (ERP) systems. Full life cycle validation is required.

Category 5: Custom (bespoke) Software
These packages are developed to meet specific needs of the user. Full life cycle validation is required and supplier audits are usually appropriate.

Figure 11.12 Categorisation of software systems according to Good Automated Manufacturing Practice guidelines (GAMP4).

requires that each operator must have a unique password to access the system at an appropriate authority level and input data or to digitally sign a record. Such records should be date-time stamped by the system, as should any subsequent alteration of a record. The systems must also satisfy data integrity and backup requirements of the regulations. The objective is to establish the same level of confidence in an electronic record as exists for a paper record – that is, to establish who did what, and when. The validation testing of such systems involves challenges to show that records cannot be created or altered without the correct password identification of the person interacting with the system, and that an audit trail of all interactions is maintained.

The validation of computer systems is now a core activity for pharmaceutical manufacturers, as most will use Enterprise Resource Planning (ERP) systems to plan manage and control their operations. Such systems are intended to be significantly configured depending on the operations undertaken at the site. They may be used to issue complete work instructions, which could include calculations based on specific batch properties, to control the release of product at the various manufacturing stages and, if validated to 21CFR Part 11, they may be used to retain an electronic batch record. Even where a system is not used to maintain an electronic record, it will most likely be used to issue paper copies of work instructions and batch records for manual completion. Thus, the accuracy and reliability of all data in the system is vital as this is what a plant is likely to operate to rather than the original paper copy procedures.

11.4.4
Methods Validation

The methods used for testing at various stages in the manufacturing process must be validated to show that they are fit for their intended application. For example, a method may be capable of measuring an analyte to a high of degree of accuracy and

Table 11.4 Example of a risk analysis matrix for assessment of a computer system.

Parameter assessed	Risk Value	Weighting	Rating
Criticality			
Quality Critical – may affect patient care	2		
Business Critical – affects areas other than GMP, e.g. Accounts	1	3	3
Validation not required	0		
System Definition			
GAMP Category 4 or 5	2		
GAMP Category 2 or a function/macro utilizing a Category 3 application	1	3	6
GAMP Category 1 or Category 3 as an application –	0		
Validation not Required System Type			
Real Time Process – may affect the patient directly (Pacemaker)	2		
Process Menu, alarm, Supervisory, Data Management	1	2	2
Business Application	0		
Relative Complexity			
Menu-driven, multi-functional	2		
Task driven, small number of functions	1	2	4
Single function or if-then-else logic	0		
System Lifetime			
Under 12 months old or Over 12 months and not Validated	2		
Over 12 months old and Partially Validated	1	1	2
To be rendered redundant within 12 months	0		
Documentation Availability			
No design specifications available	2		
Documentation exists, but is not of a high standard	1	3	3
Design specifications and procedures are well documented	0		
Computerised System Validation			
No Computerised System Validation performed	2		
Some functional testing performed	1	3	6
Computerised System Validation performed	0		
Safety (Persons using the System rather than Patient Safety)			
High risk of Injury	2		
Medium Risk	1	1	1
Low Risk	0		
Cost of Downtime/Economical Ranking			
High Cost	2		
Medium Cost	1	2	0
Low Cost	0		

(Continued)

Table 11.4 (*Continued*)

Parameter assessed	Risk Value	Weighting	Rating
Quantity of Duplicate Systems (Possible problem with data Integrity where data held and duplicated in several systems)			
Greater than 5 systems	2		
Between 2 and 5 systems	1	3	0
1 system	**0**		
Environmental Impact (e.g. Of Systems controlling Sterilizing Units)			
High	2		
Medium	1	1	0
Low	**0**		

Assigned risk values in bold   Rating = Risk Value × Weighting

precision in a series of standard solutions. However, the same level of performance may not be guaranteed when it is used to measure the analyte in a reaction mixture, due to the properties of the mixture or the presence of other compounds that may interfere with the method. Thus, analytical methods will have to be appropriately qualified for the intended use in terms of specificity, linearity, measuring range, accuracy, precision, detection limit, quantitation limit and robustness. Explanations of these terms, together with their relevance to different types of analytical procedures, are shown in Figure 11.13 and Table 11.5, respectively. Further information may be found in ICH guideline Q2, which is devoted to the topic. While full validation testing is required for new methods, published pharmacopoeial methods will only require limited evaluation to demonstrate that the method has been set up correctly.

11.4.5
Cleaning Validation

The cleaning of equipment that comes in contact with the product, such as reactors, mixing vessels, pipe-work, pumps and tubing, is another operation that is vital in order to prevent contamination due to the carry-over of residues from previous products, or of microbial growth. This may involve selecting equipment and procedures for cleaning in place (CIP) with various wash solutions, in preference to disassembly and engaging in manual cleaning. One reason for choosing a CIP option is that it is likely to deliver consistent performance, in contrast to manual operations, which may be more prone to individual operator variation. Second, an absence of direct human contact and exposure to the general environment makes it easier to obtain a sterile environment. The effectiveness of all cleaning procedures must be established by validation studies. These may involve selecting combinations of products which have very tight specifications for contaminant carry-over, or are sensitive to contamination, or which are known to be particularly difficult to clean. Testing may involve taking swabs from internal surfaces to check residue levels or microbial status as appropriate.

Specificity
Specificity is the ability to assess unequivocally the analyte in the presence of components, which may be expected to be present. Typically these might include impurities, degradants, matrix, etc.

Accuracy
The accuracy of an analytical procedure expresses the closeness of agreement between the value, which is accepted either as a conventional true value or an accepted reference value and the value found. This is sometimes termed "trueness".

Precision
The precision of an analytical procedure expresses the closeness of agreement (degree of scatter) between a series of measurements obtained from multiple sampling of the same homogeneous sample under the prescribed conditions. Precision may be considered at three levels: repeatability (within run); intermediate precision (over time); and reproducibility (inter-laboratory).

Detection Limit
The detection limit of an individual analytical procedure is the lowest amount of analyte in a sample which can be detected but not necessarily quantitated as an exact value.

Quantitation Limit
The quantitation limit of an individual analytical procedure is the lowest amount of analyte in a sample, which can be quantitatively determined with suitable precision and accuracy.

Linearity
The linearity of an analytical procedure is its ability (within a given range) to obtain test results, which are directly proportional to the concentration (amount) of analyte in the sample.

Range
The range of an analytical procedure is the interval between the upper and lower concentration (amounts) of analyte in the sample (including these concentrations) for which it has been demonstrated that the analytical procedure has a suitable level of precision, accuracy and linearity.

Robustness
The robustness of an analytical procedure is a measure of its capacity to remain unaffected by small, but deliberate, variations in method parameters; this provides an indication of its reliability during normal usage.

Figure 11.13 An explanation of terms used to define the performance characteristics of an assay

11.4.6
Validation of Sterilisation Procedures

Pharmaceuticals for injection must be presented in a sterile form. Sterility may be achieved by filtration through 0.22 µm filters under aseptic conditions, or by steam, dry heat, radiation or gas sterilisation methods, which may be applied to packaged products. Irrespective of the method, the process must be validated and monitored to assure its effectiveness. As discussed in Chapter 2, this is an example of a process that cannot be assured by verification testing because of its destructive nature.

11.4.7
Water Purification System Validation

Water purification systems must be validated to demonstrate that they can produce the required water quality. The FDA have produced guidelines for commissioning a water purification system, that requires various levels of monitoring as the

Table 11.5 Applicability of performance characteristics to different analytical procedures.

Purpose of analytical procedure	Identification	Testing for impurities		Assay - dissolution (measurement only) - content/potency
Characteristics		Quantitat.	Limit	
Accuracy	−	+	−	+
Precision				
Repeatability	−	+	−	+
Interm. Precision	−	+ (1)	−	+ (1)
Specificity (2)	+	+	+	+
Detection Limit	−	− (3)	+	−
Quantitation Limit	−	+	−	−
Linearity	−	+	−	+
Range	−	+	−	+

− Signifies that this characteristic is not normally evaluated.
+ Signifies that this characteristic is normally evaluated.
(1) In cases where reproducibility has been performed, intermediate precision is not needed.
(2) Lack of specificity of one analytical procedure could be compensated by other supporting analytical procedure(s).
(3) May be needed in some cases.

commissioning process moves from an initial 2-week first phase to a second 4-week phase, after which routine monitoring is acceptable.

11.5
GMP Requirements for Devices

In Europe there are no specific GMP regulations for medical devices. However, most manufacturers will be operating to a recognised quality system such as ISO 13485:2003, 'Medical devices – Quality management systems – Requirements for regulatory purposes', or one of the more general ISO 9000 series of quality assurance standards. The FDA have issued current GMP regulations for medical devices in the form of the Quality System Regulation (QSR) that may be found in Chapter 21 of the Code of Federal Regulation, Part 820 (21 CFR 820). The QSR is based on the original ISO quality standards, although the sequence of topics no longer match as the ISO standards were revised. The topics covered by ISO13485 and the QSR are shown in Figures 11.14 and 11.15.

By virtue of the broad diversity of products covered under the medical devices umbrella, GMP regulations and standards for devices tend to be less specific than their drug GMP counterparts. However, it is expected that the manufacturer will use risk analysis to identify specific measures that are appropriate to the manufacturing process under consideration. In effect, this means that many of the GMP measures applicable to drugs will also apply to devices. For example, if a manufacturer is

1 .Scope
1.1 General
1.2 Application

2 Normative references

3 Terms and definitions

4 Quality management system
4.1 General requirements – Implement and maintain a quality system
4.2 Documentation requirements – Create quality policy, quality manual, procedures, and document control and distribution function. Retain records for the expected lifetime of the device or a minimum of 2 years

5 Management responsibility
5.1 Management commitment – Show evidence of top management commitment to quality
5.2 Customer focus – Determine and satisfy customer requirements
5.3 Quality policy – Have top management express quality intentions
5.4 Planning – Set quality objectives and plans to meet those objectives
5.5 Responsibility, authority and communication – Appoint a quality manager, define organisational structure and responsibilities, communicate quality performance
5.6 Management review – Senior management to conduct regular reviews of the quality system and its performance

6 Resource management
6.1 Provision of resources – Provide resources to implement quality system and meet regulatory and customer requirements
6.2 Human resources – Ensure appropriate qualifications, education, training of personnel
6.3 Infrastructure – Provide appropriate facilities, equipment and supporting services
6.4 Work environment – Establish hygiene, clothing and environmental conditions appropriate to operations

7 Product realisation
7.1 Planning of product realisation – Plan and develop process for product manufacture based on risk analysis, and determine verification, validation, monitoring test and inspection requirements
7.2 Customer-related processes – Review requirements before committing to supply, communicate with customers on comments or complaints
7.3 Design and development – Establish design control process (see Unit 9)
7.4 Purchasing – Qualify suppliers, establish material specifications and verify purchased product
7.5 Production and service provision – Operate production and service provision under controlled conditions, maintain batch records, validation processes that cannot be verified, maintain identification and traceability of materials, address specific requirements for sterile products, provide suitable conditions for storage and distribution
7.6 Control of monitoring and measuring devices – Maintain and calibrate measuring equipmen

8 Measurement, analysis and improvement
8.1 General – Plan and implement monitoring, measurement, analysis and improvement processe
8.2 Monitoring and measurement – Establish customer complaints/feedback and internal audits processes, evaluate effectiveness of processes, verify product meets acceptance criteria
8.3 Control of non-conforming product – Establish procedures for rework, reject, or release under concession
8.4 Analysis of data – Analyse performance of quality system based on feedback, conformity of product to requirements, trends and supplier performance
8.5 Improvement – Implement Corrective Action, Preventative Action (CAPA) procedures, incorporating root cause investigation

Figure 11.14 The main headings of the ISO 13485:2003 standard (Standard headings reproduced with permission of International Standards Organisation).

Subpart A General Provisions
Sec. 820.1 Scope – Class II & III, plus a few Class I
Sec. 820.3 Definitions
Sec. 820.5 Quality System – Establish and maintain a quality system

Subpart B Quality System Requirements
Sec. 820.20 Management responsibility – Establish quality policy and objectives, organisational structure and responsibilities, appoint a quality manager, provide adequate resources, conduct management reviews of the quality system and establish quality plans and quality system procedures, including an overview of system
Sec. 820.22 Quality audit – Conduct regular audits of the quality system
Sec. 820.26 Personnel – Ensure appropriate qualifications, education, training of personnel

Subpart C Design Controls
Sec. 820.30 Design controls – Establish design control process (see Unit 9)

Subpart D Document Controls
Sec. 820.40 Document controls – Establish procedures for document approval, revision and distribution

Subpart E Purchasing Controls
Sec. 820.50 Purchasing controls – Evaluate suppliers and establish material specifications

Subpart F Identification and Traceability
Sec. 820.60 Identification – Identify components at all stages in the production process
Sec. 820.65 Traceability – Use lot numbers for traceability of devices /components for surgical implant or sustaining life

Subpart G Production and Process Controls
Sec. 820.70 Production and process controls – Address production procedures and process controls, changes to the process, environmental controls, clothing and hygiene of personnel, prevention of contamination, suitability and layout of buildings, equipment qualification, maintenance, periodic inspection, and adjustment, removal of unwanted manufacturing materials from devices and automated (computer controlled) processes
Sec. 820.72 Inspection, measuring and test equipment – Use traceable calibration standards and maintain calibration records for measuring equipment
Sec. 820.75 Process validation – Processes that cannot be fully verified must be validated

Subpart H Acceptance Activities
Sec. 820.80 Receiving, in-process, and finished device acceptance – Establish procedures for testing in-coming goods, in-process materials and finished product and maintain records
Sec. 820.86 Acceptance status – Disposition of products shall be clearly displayed, e.g. on hold, approved, reject.

Subpart I Non-conforming Product
Sec. 820.90 Non-conforming product – Establish procedures for control on non-conforming product, including rework and review of disposition

Subpart J Corrective and Preventative Action
Sec. 820.100 Corrective and preventative action – Implement Corrective Action, Preventative Action (CAPA) procedures, incorporating root cause investigation

Subpart K Labelling and Packaging Control
Sec. 820.120 Device labelling – Ensure legibility and durability of labels, inspect labels before release from storage, and take measures to prevent mix-up
Sec. 820.130 Device packaging – Packaging must provide adequate protection of the device

Figure 11.15 Headings of the Quality System Regulations (QSR) from 21 CFR Part 820.

Subpart L Handling, Storage, Distribution and Installation

Sec. 820.140 Handling – Establish procedures to prevent mix-ups, damage or deterioration during handling

Sec. 820.150 Storage – Establish procedures for proper storage and control

Sec. 820.160 Distribution – Establish procedures for distribution and maintain distribution records

Sec. 820.170 Installation – Establish procedures for correct installation, where applicable

Subpart M Records

Sec. 820.180 General requirements – Retain records for the expected shelf life of the device or a minimum of 2 years after last distribution

Sec. 820.181 Device master record – Compile or refer to procedures and specifications for device manufacture

Sec. 820.182 Device history record – Maintain the production records for each batch of devices

Sec. 820.186 Quality system record – Compile or refer to location of general procedures relating to the quality system

Sec. 820.198 Complaint files – Maintain complaint files, investigate and assess if there are reporting obligations, keep copies of files available in the US either at manufacturer or initial distributor site

Subpart N Servicing

Sec. 820.200 Servicing – Maintain servicing procedures where applicable

Subpart O Statistical Techniques

Sec. 820.50 Statistical techniques – Use appropriate statistical methods for data analysis and sampling plans

Figure 11.15 (*Continued*)

producing an implantable device, then the same levels of cleanliness and sterility as are required for an injectable biopharmaceutical will apply. Thus, device manufacturers will often use drug GMP guidance as a basis for developing appropriate manufacturing conditions for devices. It is worthwhile noting that, just as US drug GMPs require the compilation of a master production and control record, US device regulations require the preparation of an equivalent Device Master Record (DMR), which should contain or reference all the procedures and specifications necessary for its manufacture.

11.6
Chapter Review

Good manufacturing practices have been developed to ensure that drugs and devices are manufactured to the requisite standards of cleanliness, integrity and quality. These are drawn from the general principles of quality management systems, which have been extended to address specific aspects that are necessary to ensure the quality of such products. GMP requirements generally increase as product reaches the final stages in the manufacturing process. The validation of facilities, equipment, systems and processes is central to establishing GMP operations.

11.7
Further Reading

- EU GMP Regulations
  Directive 2003/94/EC laying down the principles and guidelines of good manufacturing practice in respect of medicinal products for human use and investigational medicinal products for human use.
  Directive 91/412/EEC laying down the principles and guidelines of good manufacturing practice for veterinary medicinal products.
  The Rules Governing Medicinal Products in the European Union, Volume 1, Legislation Human, Volume 5, Legislation Veterinary. http://ec.europa.eu/enterprise/pharmaceuticals/eudralex/index.htm.
- EU GMP Guidance
  The Rules Governing Medicinal Products in the European Union, Volume 4, Medicinal Products for Human and Veterinary Use: Good Manufacturing Practice.
  http://ec.europa.eu/enterprise/pharmaceuticals/eudralex/index.htm.
- US GMP Regulations
  21 CFR Part 210, current Good Manufacturing Practice in Manufacturing, Processing, Packing, or Holding of Drugs, General Requirements.
  21 CFR Part 211, Current Good Manufacturing Practice for Finished Pharmaceuticals.
  21 CFR Part 820 Quality System Regulation Medical Devices; Current Good Manufacturing Practice.
  www.fda.gov.
- Harmonised GMP Guidelines
  ICH Guideline: Q7 Good Manufacturing Practice for Active Pharmaceutical Ingredients. www.ich.org.
- Process Validation
  Globally Harmonised Task Force Guidance: Quality Management Systems - Process Validation.
  www.ghtf.org.
  FDA Guidance: Guideline on general principles of process validation.
  www.fda.gov.
- Methods Validation
  ICH Guideline: Q2 Validation of Analytical Procedures: Text and Methodology.
  www.ich.org.
- Computer System Validation
  US Code of Federal Regulations 21 CFR Part 11.
  FDA Guidance: Guidance for Industry Part 11, Electronic Records; Electronic Signatures – Scope and Application.
  www.fda.gov.

12
Oversight and Vigilance

12.1
Chapter Introduction

The previous chapter outlined the Good Manufacturing Practice (GMP) principles that a manufacturer must apply in order to meet the required standards of quality, identity and purity. This chapter will look at the regulatory requirements regarding oversight of manufacturing operations and product performance in the marketplace. The key objectives of these requirements are to ensure that only consistently high-quality products are produced, and to facilitate the early detection of problems in the marketplace. Manufacturing operations are controlled via requirements for the registration and licensing of manufacturers and the conduct of site inspections to guarantee that appropriate GMPs are applied. The most stringent regulatory controls are applied to the manufacture of finished pharmaceutical products. Obligations as regards market vigilance and other reporting duties, allows regulatory authorities to be kept informed of potential issues in the marketplace and to take appropriate action where necessary.

12.2
Registration of Manufacturers and Other Entities

As a prelude to exercising regulatory oversight of manufacturing operations, regulatory authorities must be made aware of the entities involved. Different procedures and requirements apply, depending on the type of product and the applicable regulatory regime.

12.3
Manufacturing Authorisation of Medicinal Products in the EU

According to Article 40 of Directive 2001/83/EC (human medicinal products) and Article 44 of Directive 2001/82/EC (veterinary medicinal products), pharmaceutical

Medical Product Regulatory Affairs. John J. Tobin and Gary Walsh
Copyright © 2008 WILEY-VCH Verlag GmbH & Co. KGaA, Weinheim
ISBN: 978-3-527-31877-3

companies may not manufacture, package or label medicinal products in the EU without obtaining a manufacturing authorisation. The authorisation requirement also applies to companies that import pharmaceuticals into the EU from non-EU countries. A separate authorisation (licence) is required for each site where such activities take place. To obtain a licence, an application form must be submitted to the Competent Authority of the Member State in whose territory the site is located. The application must be accompanied by documented information on the following:

- The products manufactured or controlled at the site.
- The identity and qualifications of the person responsible for product release (a "Qualified Person").
- A description of the facilities equipment and quality control arrangements at the site.

EU regulations require that each facility has at its disposal at least one "Qualified Person", who is responsible for the review and release of finished medicinal products. The regulations specify various combinations of education, training and experience that may be presented in order for a person to be recognised as a Qualified Person. Many Qualified Persons hold degrees in medicine, veterinary science or pharmacy. Before releasing a batch, the Qualified Person must review all the batch paperwork and Quality Control (QC) test records to ensure that the product has been manufactured and checked in accordance with GMP regulations and the requirements of the marketing authorisation. For products manufactured outside the EU, product release will be based on a review of paperwork supplied by the manufacturer coupled with testing performed at the EU site, unless there is an agreement on mutual recognition of GMP inspections conducted by non-EU states.

Information on the site facilities and operations should be presented in the form of a Site Master File (SMF). A Pharmaceutical Inspection Cooperation Scheme (PICS) committee has developed specific guidance on the preparation of a SMF, the suggested content of which is outlined in Figure 12.1. The purpose of the SMF is to provide the Competent Authorities with a basic understanding and outline of activities at the site so as to enable them to effectively plan GMP inspections. The Competent Authority must conduct a GMP inspection as part of the licensing process. The authorities are allowed 90 days to issue a licence, which is specific for both the products and the activities handled at the site. Changes to either must be authorised by the authorities. The time allowed to authorise a change is 30 days. It should be noted that the manufacture or import of products for human clinical trials are subject to the same authorisation procedures. A central database of manufacturing authorisations is maintained by the European Medicines Agency (EMEA).

12.3.1
Wholesale Distribution of Medicinal Products

Sites that warehouse and engage in the wholesale distribution of medicinal products are also subject to similar authorisation procedures. The sites must have suitable premises to properly store and preserve the products, employ a designated qualified

1   GENERAL INFORMATION

1.1   Brief information on the firm (including name and address), relation to other sites and, particularly, any information relevant to understand the manufacturing operations.

1.2   Pharmaceutical manufacturing activities as licensed by the Competent Authorities.

1.3   Any other manufacturing activities carried out on the site.

1.4   Name and exact address of the site, including telephone, fax and 24-hour telephone numbers.

1.5   Type of actual products manufactured on the site, and information about specifically toxic or hazardous substances handled, mentioning the way they are manufactured (in dedicated facilities or on a campaign basis).

1.6   Short description of the site (size, location and immediate environment and other manufacturing activities on the site).

1.7.   Number of employees engaged in the quality assurance, production, quality control, storage and distribution.

1.8   Use of outside scientific, analytical or other technical assistance in relation to manufacture and analysis.

1.9   Short description of the quality management system of the firm responsible for manufacture.

2   PERSONNEL

2.1   Organisation chart showing the arrangements for quality assurance, including production and quality control.

2.2   Qualifications, experience and responsibilities of key personnel.

2.3   Outline of arrangements for basic and in-service training and how records are maintained.

2.4   Health requirements for personnel engaged in production.

2.5   Personnel hygiene requirements, including clothing.

3   PREMISES AND EQUIPMENT

Premises

3.1   Simple plan or description of manufacturing areas with indication of scale (architectural or engineering drawings are not required).

3.2   Nature of construction and finishes.

3.3   Brief description of ventilation systems. More details should be given for critical areas with potential risks of airborne contamination (schematic drawings of the systems are desirable). Classification of the rooms used for the manufacture of sterile products should be mentioned.

3.4   Special areas for the handling of highly toxic, hazardous and sensitising materials.

3.5   Brief description of water systems (schematic drawings of the systems are desirable) including sanitation.

3.6   Maintenance (description of planned preventive maintenance programmes and recording system).

Equipment

3.7   Brief description of major production and control laboratories equipment (a list of equipment is not required).

3.8   Maintenance (description of planned preventative maintenance programmes and recording system).

Figure 12.1 The contents of a Site Master File. Extract from "PIC/S Explanatory Notes For Industry on the Preparation of a Site Master File" (PE008) © PIC/S September 2007.

3.9   Qualification and calibration, including recording system. Arrangements for computerised systems validation.

Sanitation

3.10 Availability of written specifications and procedures for cleaning manufacturing areas and equipment.

4   DOCUMENTATION

4.1   Arrangements for the preparation, revision and distribution of necessary documentation for manufacture.

4.2   Any other documentation related to product quality, which is not mentioned elsewhere (e.g. microbiological controls on air and water).

5   PRODUCTION

5.1   Brief description of production operations using, wherever possible, flow sheets and charts specifying important parameters.

5.2   Arrangements for the handling of starting materials. Packaging materials, bulk and finished products, including sampling, quarantine, release and storage.

5.3   Arrangements for reprocessing or rework.

5.5   Brief description of general policy for process validation.

6   QUALITY CONTROL

6.1   Description of the Quality Control system and of the activities of the Quality Control Department Procedures for the release of finished products.

7   CONTRACT MANUFACTURE AND ANALYSIS

7.1   Description of the way in which the GMP compliance of the contract acceptor is assessed.

8   DISTRIBUTION, COMPLAINTS AND PRODUCT RECALL

8.1   Arrangements and recording system for distribution.

8.2   Arrangements for the handling of complaints and product recalls.

9   SELF-INSPECTION

9.1   Short description of the self-inspection system.

Figure 12.1 (*Continued*)

person with responsibility for ensuring that activities are carried out in compliance with the regulations, distribute authorised products only to other authorised distributors or appropriate sales outlets (pharmacies, licensed merchants, medical practitioners, veterinarians, retail outlets, as appropriate to the product category) maintain records of distribution, and maintain an emergency plan to facilitate effective recall of product from the market. The authorisation timelines are the same as for manufacturing sites, and similarly Competent Authorities are required to inspect the sites prior to issuing a licence.

12.3.2

Registration of Persons Responsible for Placing Medical Devices on the EU Market

As described previously, persons responsible for placing medical devices on the EU market are only required to notify the Competent Authority in the state where their business is registered of the name and address of the business, together with a description of the devices marketed. In the case of a multi-site or multinational organisation, this may be the corporate headquarters address from which marketing operations are carried out. This is similar to marketing authorisations for drugs, where the Marketing Authorisation Holder (MAH) or Product Authorisation (PA) Holder is often in the name and address of the marketing headquarters. The registered person is responsible for maintaining the technical documentation that supports the application of the CE mark of conformity available for inspection by the Competent Authorities. If a device manufacturer does not have a presence in any EU Member State, then they must use the services of an authorised representative, termed an "EU Rep", for this purpose. Distinct from drug manufacturers, there is no requirement to register individual manufacturing sites, as the Competent Authorities are not legally obliged to conduct inspections of such sites. Instead, such sites are likely to be inspected by Notified Bodies as part of the process of maintaining certificates of compliance with recognised quality systems.

12.4
Registration of Producers of Drugs and Devices in the US

According to Section 510 of the Food Drug and Cosmetic Act, producers of drugs and devices are required to register the name and address of each establishment with the FDA. These include all US-based sites involved in manufacturing, packaging, labelling, or testing of drugs, drug components or devices, as well as foreign-based manufacturers that export products to the US. Wholesale distributors of drugs must also register, but device distributors are exempt from this requirement. In addition to registering their establishment, the registrants are also required to submit product listings accompanied by the product labelling.

Detailed regulations concerning the registration of establishments and product listings are set out in 21 CFR Parts 207 (drugs), 21 CFR Parts 607 (blood/blood products) and 21 CFR Parts 807 (devices) and 9 CFR Part 102 (veterinary biologics). As summarised in Table 12.1, specific forms have been created for notification, depending on the type of product involved. An example of the registration form for a drugs facility is shown in Figure 12.2. Establishment registration forms must be submitted at initial start up and annually thereafter. Additionally, a foreign registrant must identify a US-based agent and should also list known importers. Product listings may be submitted twice annually to cover any material changes from the previous listing (new products, discontinued products, modifications). Registration is mainly an administrative exercise and will result in the issuance Establishment Registration Numbers and product listing codes. Unlike the manufacturing authorisation process in Europe, establishment registration is not directly coupled with a site inspection. However, registration means that the site will be included in the database of sites that are subject to a bi-annual inspection regime to ensure that they comply with GMP and other requirements.

Table 12.1 Forms used for establishment and product listings.

Form	Purpose	Submit to
FDA-2656	Registration of Drug Establishment	CDER
FDA-2657	Drug Product Listing	CDER
FDA-2830	Blood Establishment Registration and Product Listing	CBER
FDA-2891	Initial Registration of Device Establishments	CDRH
FDA-2891a	Registration of Device Establishment	CDRH
FDA-2892	Medical Device Listing	CDRH
FDA-3356	Human tissues and cell Establishment Registration and Product Listing	CBER

12.4.1
Additional Licensing Requirements

Many sites will have to fulfil obligations under relevant environmental and safety legislation. Sites that handle chemicals or engage in operations that are potentially dangerous to the environment will have to register with the relevant environmental protection agency. Sites will be inspected to ensure that there are adequate controls, containment measures and emergency response procedures in place to prevent damage to the environment. Integrated pollution prevention and control licenses will be issued, which specify the conditions and restrictions under which the facility may operate. Site which hold quantities of hazardous chemicals above threshold levels that could result in major accidents due to explosion, fire or toxic releases, must register with the local authorities. They must prepare emergency plans that are integrated with those of the local emergency services.

12.5
Inspections

The regulatory authorities must conduct regular (usually every 2 years) GMP inspections of sites in their territory that are subject to GMP regulations. These are normally scheduled visits, but could be un-announced if the circumstances so warrant. Active pharmaceutical ingredient (API) manufacturing sites are included as part of the routine inspection programme under US regulations. In Europe, the authorities may inspect API manufacturing sites or licence holder's premises where it is deemed necessary. Foreign manufacturing sites must permit inspections by authorities from states where the product is being marketed. However, in practice such foreign inspections have been greatly reduced by the establishment of mutual recognition agreements where the parties accept the validity of their respective GMP inspection regimes. The European Commission negotiates such agreements collectively on behalf of the Member States. GMP mutual recognition agreements have been concluded with the USA, Canada, Australia, New Zealand, Switzerland and Japan, among others.

Form Approved: OMB No. 0910-0045. Expiration Date: December 31, 2007.   See OMB Statement on Reverse.

DEPARTMENT OF HEALTH AND HUMAN SERVICES FOOD AND DRUG ADMINISTRATION **REGISTRATION OF DRUG ESTABLISHMENT/ LABELER CODE ASSIGNMENT** (In accordance with Public Law 92-387)	FDA USE ONLY		FDA USE ONLY	

NOTICE: This report is required by law (21 C.F.R. 207.20). Failure to report can result in imprisonment for not more than one year or a fine of not more than $1,000, or both. (FD&C Act, Section 303).

	LABELER CODE	REGISTRATION NUMBER

SECTION A - SITE INFORMATION

REPORTING FIRM NAME — STATE OF INC.

SITE ADDRESS (No P.O. Box) — SITE TELEPHONE NUMBER ( )

CITY	STATE	ZIP CODE	COUNTRY	BUSINESS CATEGORY
				☐ HUMAN ☐ VETERINARY

SITE MAILING ADDRESS (If different from site address)

CITY	STATE	ZIP CODE	COUNTRY	SITE INTERNET/EMAIL ADDRESS

DOING BUSINESS AS (DBA) NAME OF FIRM (if applicable)

PARENT COMPANY NAME

REASON(s) FOR SUBMISSION	TYPE OF OWNERSHIP	PERSON SUBMITTING DATA AND TELEPHONE
☐ Firm Registration ☐ Address Change	☐ Sole Proprietorship	
☐ Registration of ☐ Merger/Buyout	☐ Partnership	**BUSINESS TYPE**
☐ Additional Site ☐ Reentry into Business	☐ Coop. Assn.	☐ Distributor*
☐ Re-Registration with Same Name	☐ Corporation	☐ Manufacturer ☐ Foreign Country
☐ LC Assignment ☐ Out of Business	☐ Other____	☐ Repacker ☐ Analytical Lab
☐ Name Change		☐ Relabeler ☐ Other____

SECTION B - FIRM COMPLIANCE MAILING ADDRESS for Annual Listing Report and/or Firm Correspondence

NUMBER AND STREET AND/OR P.O. BOX and ATTENTION LINE and/or Internal Mail Code — TELEPHONE NUMBER ( )

CITY	STATE	ZIP CODE	COUNTRY	COMPLIANCE INTERNET/EMAIL ADDRESS

SECTION C - ADDITIONAL FIRM AND SITE INFORMATION

NAME OF OWNER, PARTNERS OR OFFICERS	TITLE	POSITION

OTHER FIRMS DOING BUSINESS AT THIS SITE

LABELER CODE	FIRM NAME	LABELER CODE	FIRM NAME

SECTION D - SIGNATURE

SIGNATURE OF AUTHORIZING OFFICIAL	TITLE	DATE

*DISTRIBUTOR'S CERTIFICATION: As a, Distributor, I am submitting product listing information to the FDA on my own behalf. I have provided a copy of this certification (Form FDA 2656) to the registered manufacturer(s). My signature and phone number are listed below.

RETURN THIS FORM TO:	SIGNATURE OF DISTRIBUTOR
FOOD AND DRUG ADMINISTRATION	
CDER DRUG REGISTRATION AND LISTING, (HFD-143)	
5901 B AMMENDALE ROAD	DISTRIBUTOR'S TELEPHONE NUMBER
BELTSVILLE, MD 20705	
INTERNET: DRUGLISTING@CDER.FDA.GOV	( )

FDA 2656 (4/06)   NOTE: Validation of this form is not to be construed as FDA approval of the establishment or its products.
PREVIOUS EDITION IS OBSOLETE

PSC Graphics (301)-443-1090 EF

Figure 12.2 Form FDA 2656 used for registration of a drug establishment.

If using <u>Federal Express, DHL or any special carrier</u> to return the forms, please use the following address:

(Please refer to the Drug Registration and Listing Instruction Booklet.)

When completing this form, please refer to the Drug Registration and Listing Instruction Booklet for assistance.
PLEASE PRINT IN ENGLISH USING BLACK INK.

Public reporting burden for this collection of information is estimated to average 30 minutes per response, including the time for reviewing instructions, searching existing data sources, gathering and maintaining the data needed, and completing and reviewing the collection of information. Send comments regarding this burden estimate or any other aspect of this collection of information, including suggestions for reducing this burden to:

Food and Drug Administration
Information Management Team, HFD-095
5600 Fishers Lane
Rockville, MD 20857

An agency may not conduct or sponsor, and a person is not required to respond to, a collection of information unless it displays a currently valid OMB control number.

FORM FDA 2656 (4/06) (BACK)

Figure 12.2 (*Continued*)

It has not been possible to include devices in the GMP agreement between Europe and the US. Under US regulations, bi-annual GMP inspections of device manufacturers must be conducted to ensure compliance with the Quality System Regulation. Traditionally, the FDA as the regulatory authority has been responsible for these, although recent reform of the regulations permits accredited third-party bodies to inspect sites producing class II or lower devices. In Europe, regular inspections of device-manufacturing sites are conducted by Notified Bodies for the purposes of certifying compliance with relevant ISO quality standards. The regulatory authorities (Competent Authorities) may visit occasionally for monitoring purposes, or in response to compliance issues that may have surfaced through market surveillance; however, there is no legal obligation on them to routinely conduct or certify inspections. Consequently, European device manufacturers that export to the US will receive FDA inspections, although usually not at the same frequency as would apply to US-based manufacturers.

12.5.1
Inspection Techniques

Irrespective of the particular focus of an inspection or the auditing strategy adopted, most audits will have two basic elements:

- A review of documentation and procedures to verify that they comply with the relevant requirements.
- Checks to ensure that the conditions and practices outlined in the documentation are actually achieved.

In the case of pharmaceutical sites, inspectors will have the opportunity to do some of the documentation verification in advance of a site visit. Relevant information will be available either as a SMF submitted with a manufacturing authorisation application in Europe, or the chemistry manufacturing and control section of a drug marketing application, which will have been forwarded to the local field office, in the case of FDA inspections. Actual site audits commence with a meeting where quality management and usually other senior managers, are briefed as regards the general outline and schedule. The inspector may then opt for a quick orientation tour or else proceed with a document review. Having identified what is supposed to happen, the second stage is to verify that this in fact takes place. This may be accomplished by a combination of record reviews and inspection tours. Inspection tours are usually the most difficult part of an audit to "manage" from a manufacturer's perspective, as the inspector will be observing actual operations and conditions and will frequently talk to staff to check that they have sufficient knowledge of the task that they are undertaking and its impact on GMP. Inspectors may also remove samples of materials for independent testing.

The nature of an inspection will vary depending on the purpose of the audit and the type and size of the facility under scrutiny. Initial audits of a new facility will usually take the longest, as inspectors will want to conduct a full inspection of all aspects of the operation. Subsequent routine surveillance audits will tend to be shorter as auditors may just cover selected elements. Usually, this would include areas of concern from previous audits, and in this regard it is vital that any such issues are rectified when preparing for the next audit. It is also wise to ensure that any issues that may have arisen in the market are adequately addressed, as these are likely to be checked next time around. Serious adverse events on the market may trigger an immediate "for cause" inspection if it raises doubts as to the effectiveness of, or the commitment to, the GMP system. It is worthwhile to note that if an identified deficiency can be corrected during the course of an audit, it is prudent to do so as this underlines the site's commitment to compliance.

Inspectors from different jurisdictions may adopt different inspection strategies. The Pharmaceutical Inspection Cooperation Scheme (PICS) provides training for inspectors with a view to achieving greater harmony in the standards of inspection among their member authorities. They organise conferences, which discuss various aspects of GMP, and also arrange multinational inspection teams, which enable inspectors from different countries to learn from each other. Direction is also available from the EMEA, who have compiled a set of guidance documents for GMP inspections. European inspections will tend to be conducted as either general GMP inspections when monitoring for routine purposes, or as product- or process-oriented inspections in response to specific compliance issues (e.g. re-audits, or product quality problems). Inspectors will tend to report findings under the same headings as used for European GMP regulations, namely:

- Quality Management
- Personnel
- Premises and Equipment
- Documentation
- Production
- Quality Control
- Contract Manufacture and Analysis
- Complaints and Product Recall
- Self Inspection

The FDA have developed a systems approach for audit purposes. They have grouped the GMP requirements for pharmaceutical sites into six major systems, as outlined in Figure 12.3. A full GMP inspection of a pharmaceutical site, which would be conducted for a new site, or an existing site where there are serious concerns as regards the level of compliance, will cover all six systems. Abbreviated audits are normally employed for routine surveillance monitoring. These must include the quality system, which is regarded as the core system that knits all the other systems together, plus one other system. The choice of second system may be influenced by aspects such as the introduction of significant changes or the desire to cover all systems over a series of audits. An abbreviated audit may also be used for issue specific compliance audits.

In 2002, the FDA announced a significant new initiative, "Pharmaceutical Current Good Manufacturing Practices (cGMPs) for the 21st Century". The purposes of this

1. Quality System.
This system assures overall compliance with cGMPs and internal procedures and specifications. The system includes the quality control unit and all of its review and approval duties (e.g. change control, reprocessing, batch release, annual record review, validation protocols, and reports, etc.). It includes all product defect evaluations and evaluation of returned and salvaged drug products.

2. Facilities and Equipment System.
This system includes the measures and activities which provide an appropriate physical environment and resources used in the production of the drugs or drug products. It includes:
a) Buildings and facilities along with maintenance;
b) Equipment qualifications (installation and operation); equipment calibration and preventive maintenance; and cleaning and validation of cleaning processes as appropriate. Process performance qualification will be evaluated as part of the inspection of the overall process validation which is done within the system where the process is employed; and,
c) Utilities that are not intended to be incorporated into the product such as HVAC, compressed gases, steam and water systems.

3. Materials System.
This system includes measures and activities to control finished products, components, including water or gases that are incorporated into the product, containers and closures. It includes validation of computerised inventory control processes, drug storage, distribution controls, and records.

4. Production System.
This system includes measures and activities to control the manufacture of drugs and drug products including batch compounding, dosage form production, in-process sampling and testing, and process validation. It also includes establishing, following, and documenting performance of approved manufacturing procedures.

5. Packaging and Labelling System.
This system includes measures and activities that control the packaging and labelling of drugs and drug products. It includes written procedures, label examination and usage, label storage and issuance, packaging and labelling operations controls, and validation of these operations.

6. Laboratory Control System.
This system includes measures and activities related to laboratory procedures, testing, analytical methods development and validation or verification, and the stability program.

Figure 12.3 FDA systems approach to GMP audits of pharmaceutical sites.

are to enhance and modernize the regulation of pharmaceutical manufacturing and product quality, and bring a 21st century focus to this critical FDA responsibility. The "current" in current GMP has often been a source of debate. Although the base regulations may have remained unchanged for some time, there is an expectation that manufacturers update their practices to reflect technological and scientific advances. This is further reinforced by the new approach, as it is included among the objectives, which are stated as follows:

• To encourage the early adoption of new technological advances by the pharmaceutical industry.

• To facilitate industry application of modern quality management techniques, including the implementation of quality systems approaches, to all aspects of pharmaceutical production and quality assurance.

- To encourage the implementation of risk-based approaches that focus both industry and Agency attention on critical areas.

- To ensure that regulatory review, compliance, and inspection policies are based on state-of the-art pharmaceutical science.

- To enhance the consistency and coordination of the FDA's drug quality regulatory programs, in part, by further integrating enhanced quality systems approaches into the Agency's business processes and regulatory policies concerning review and inspection activities.

As this strategy evolves over time, it can be expected that the FDA will devote more of its resources to the oversight of high-risk operations and emerging technologies, with consequently less attention on low-risk activities.

The FDA also use a systems approach when auditing device manufacturing sites. This breaks down the GMP requirements for devices into seven primary subsystems consisting of:

- Management Controls
- Design Controls
- Production and Process Controls
- Records/Documents/Change Controls
- Material Controls
- Facility and Equipment Controls
- Corrective and Preventive Actions

The relationship between the main subsystems and other minor systems is illustrated schematically in Figure 12.4. This places management at the core of the quality system, with the other systems arranged as major and minor satellites that revolve around it. This perspective provides the basis for the Quality System Inspection Technique (QSIT), which the FDA uses for auditing medical device facilities. This is based on a "top-down" approach, which starts with management controls and then looks at three other key subsystems of Design Controls, Corrective and Preventative Actions (CAPA) and Production and Process Controls. The belief is that by focussing on just these four subsystems, you will actually touch on all the other subsystems and obtain a sufficiently satisfactory overview of the state of compliance of the facility.

12.5.2
Audit Findings and Consequences

On completion of an audit and before the auditors leave the premises, the management of the site will be briefed as regards the main observations during the audit. Where issues arise, FDA inspectors use Form 483 to present the manufacturer with a list items that may be deemed as non-compliant with the regulations. Should any non-conformances be highlighted at the audit close-out session, management should state their commitment to correcting these and, if possible, indicate what the

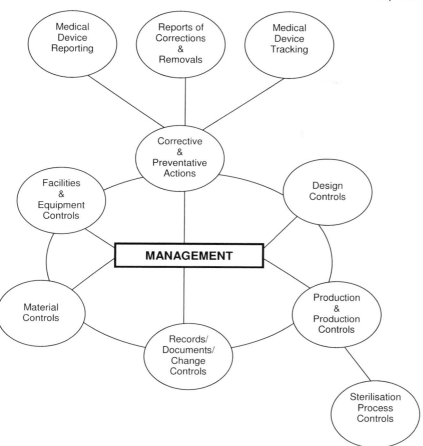

Figure 12.4 Quality System Regulation subsystems from the FDA guide to inspection of quality systems.

corrective action strategy may be. Once the audit is over, it is prudent to start to prepare a response to the auditing body outlining the proposed corrective, without waiting for an official report. Official reports are only written after leaving the site, which allows inspectors the opportunity of consulting with colleagues or obtaining the results of sample testing. An example of the report template used for European GMP audits is shown in Figure 12.5. Deficiencies may be ranked as critical, major, other, or minor, depending on the assessment scheme and the consequences for the product. Auditors may also choose to make observations with the intention of highlighting aspects that are not technically in breach of the requirements, but which could be improved upon. Manufacturers should respond immediately to the issues raised, and outline how they will be addressed. In some instances where a major overhaul of a system is required, it may be necessary to enter into negotiations on an agreed plan and timeline. If the deficiencies are viewed as serious, a follow-up audit is likely to ensue to confirm the effectiveness of corrective action. In most instances where the deficiencies are minor, they do not compromise product safety, and the manufacturer has committed to rectifying the problem, then no further action will be taken. However, if the regulatory authorities form the opinion that the manufacturer has not provided an adequate response to the issues raised, then further action will follow. Under the US system, this could escalate from an FDA warning letter to judicial action. If successful, this could result in the courts issuing of a "consent decree", penalising the company for each day it is not in compliance. Some

GMP Inspection report - Community format

Report Reference no.:	
Inspected site(s):	*Name and full address of the inspected site.*
Activities Carried out:	*Manufacture of active substance* ☐ *Manufacture of finished product* ☐ *Manufacture of intermediate or bulk* ☐ *Packaging only* ☐ *Importing* ☐ *Laboratory testing* ☐ *Batch control and batch release* ☐ *Storage and distribution* ☐ *Investigational medicinal products* ☐ *Other* _____
Inspection date(s):	*Date(s), month, year.*
Inspector(s):	*Name(s) of the inspector(s).* *Name(s) of expert / assessor (if applicable).* *Name(s) of the Competent Authority(ies).*
References:	*Reference number of Marketing and / or Manufacturing Authorisations* *EMEA reference number(s).(If the inspection is an EMEA inspection).*
Introduction:	*Short description of the company and the activities of the company.* *For inspections in non-EEA countries, it should be stated whether the Competent Authority of the country, where the inspection took place, was informed of the inspection and whether the Competent Authority took part in the inspection.* *Date of previous inspection.* *Name(s) of Inspector(s) involved in previous inspection.* *Major changes since the previous inspection.*

Figure 12.5 GMP report template (reproduced with permission of EMEA).

Brief report of the inspection activities undertaken:	
Scope of Inspection:	*Short description of the inspection (Product related, process related inspection and/or General GMP inspection, reference to specific dosage forms where appropriate). The reason for the inspection should be specified (e.g. new marketing application, routine, investigation of product defect.*
Inspected area(s):	*Each inspected area should be specified.*
Activities not inspected:	*Where necessary attention should be drawn to areas or activities not subject to inspection on this occasion.*
Personnel met during the inspection:	*The names and titles of key personnel met should be specified (listed in annex).*
Inspectors findings and observations relevant to the inspection; and deficiencies:	*Relevant headings from The Rules Governing Medicinal Products in the European Community, Good Manufacturing Practice for Medicinal Products Vol. IV. (Guide to GMP, Basic Requirements, relevant for scope of inspection).*
	This section can link the findings to the deficiencies and be used to explain classification.
	The detail in the narrative of this section of the report may be reduced where a Site Master File acceptable to the reporting authority has been submitted to the Competent Authority.
Headings to be used *New headings may be introduced when relevant*	Overview of inspection findings from last inspection and the corrective action taken.
	Quality Management
	Personnel
	Premises and Equipment
	Documentation
	Production
	Quality Control
	Contract Manufacture and Analysis
	Complaints and Product Recall
	Self Inspection
Distribution and Shipment:	e.g. Compliance with Good Distribution Practice
Questions raised relating to the assessment of a marketing application:	e.g. Pre-authorisation Inspections
Other specific issues identified:	e.g. Relevant future changes announced by company
Site Master File:	Assessment of SMF if any; date of SMF

Figure 12.5 (*Continued*)

large companies, to their cost, have ended up in this situation when they decided to challenge the FDA on their interpretation of the FDC Act. If GMP violations were to compromise public safety, then product may be seized or blocked from the market. In Europe, the auditing bodies have the option of withdrawing certificates of compliance

Miscellaneous:	
Samples taken	
Annexes attached:	*List of any annexes attached*
List of Deficiencies classified into critical, major and others:	*All deficiencies should be listed and the relevant reference to the EU GMP Guide and other relevant EU Guidelines should be mentioned.*
	All deficiencies found should be listed even if corrective action has taken place straight away.
	If the deficiencies are related to the assessment of the marketing application it should be clearly stated.
	The company should be asked to inform the Inspectorate about the proposed time schedule for corrections and on progress.
Recommendations:	*To the Committee requesting the inspection or to the Competent / Enforcement Authority for the site inspected.*
Summary and conclusions:	*The Inspector(s) should state whether, within the scope of the inspection, the company operates in accordance with the EU GMP Rules provided, where relevant, that appropriate corrective actions are implemented and mention any other item to alert requesting authority. Reference may be made to conclusions recorded in other documents, such as the close-out letter, depending on national procedures.*
Name(s):	*The inspection report should be signed and dated by the inspector(s)/assessors having participated in the inspection.*
Signatures(s):	
Organisation(s):	
Date:	
Distribution of Report:	

Figure 12.5 (*Continued*)

and manufacturing authorisations, which would in effect prevent the product from being placed on the market. A database of pharmaceutical manufacturing author- isations and GMP certificates issued by national Competent Authorities is main- tained by the EMEA on the EudraGMP network, enabling the status of an manufacturer to be easily checked. In general, punitive actions are not commonplace

2. Definition of Significant Deficiencies

2.1 Critical Deficiency:

A deficiency which has produced, or leads to a significant risk of producing either a product which is harmful to the human or veterinary patient or a product which could result in a harmful residue in a food producing animal.

2.2 Major Deficiency:

A non-critical deficiency:

which has produced or may produce a product, which does not comply with its marketing authorisation;

or

which indicates a major deviation from EU Good Manufacturing Practice;

or

(within EU) which indicates a major deviation from the terms of the manufacturing authorisation;

or

which indicates a failure to carry out satisfactory procedures for release of batches or (within EU) a failure of the Qualified Person to fulfil his legal duties;

or

a combination of several "other" deficiencies, none of which on their own may be major, but which may together represent a major deficiency and should be explained and reported as such;

2.3 Other Deficiency:

A deficiency, which cannot be classified as either critical or major, but which indicates a departure from good manufacturing practice.

(A deficiency may be "other" either because it is judged as minor, or because there is insufficient information to classify it as a major or critical).

Figure 12.5 (*Continued*)

as regulatory authorities prefer to adopt a cooperative approach with manufacturers to get them back in compliance.

12.6
Market Vigilance and Oversight of Drugs

Despite the extensive investigations that must be undertaken prior to placing a drug on the market, it is not possible to guarantee that all safety issues have been identified. Thus, market vigilance systems must be maintained after a drug has been launched so as to detect safety issues that were not evident prior to commercialisation. For such systems to be effective requires the participation and cooperation of the medical profession, the pharmaceutical industry and the regulators, in order that critical safety information can be identified and acted on in a timely manner. In some instances this has resulted in the withdrawal or restriction of drugs due to safety concerns that only emerged post-commercialisation.

12.6.1
Pharmacovigilance in the EU

The requirements for pharmacovigilance systems in Europe are contained in Directive 2001/82/EC and Directive 2001/83/EC, and Regulation (EC) No. 726/ 2004, and supported by guidance extensive guidance set out in The Rules Governing Medicinal Products in the European Union, Volume 9, Pharmacovigilance.

12.6.2
Qualified Person

In the EU, a basic requirement for marketing authorisation holders (MAHs) is that they must have permanently and continuously at their disposal a nominated qualified person, with responsibility for pharmacovigilance. This person should have experience in all aspects of pharmacovigilance and, if not the holder of a medical or veterinary qualification, they should have access to such a person. The duties of the qualified person include:

- Establishing and maintaining a pharmacovigilance system to capture and evaluate all reports of suspected adverse drug reactions.
- Preparing relevant reports on such matters for the Competent Authorities.
- Responding to requests for further information from the Competent Authorities.
- Providing the Competent Authorities with other information relevant to maintaining an up-to-date picture of the risk benefit profile of the product.

12.6.3
Reporting Requirements

Pharmacovigilance is reliant on the medical profession as the primary source of information on suspected adverse drug reactions. Systems must be in place to facilitate reporting by medical personnel directly to their national Competent Authority, as well as to the drug supplier. Other sources of information that the marketing authorisation holder is required to take into consideration, includes reports from non EU markets, the published scientific literature, and data from post-authorisation safety studies. Such MAH-sponsored safety studies are initiated, usually in consultation with the regulatory authorities, with the purpose of obtaining more information on safety aspects under the normal usage patters, particularly where this involves long-term use or there is a potential safety concern that requires further investigation. Many safety studies will generally focus on collecting and analysing observational data under the normal prescribing regime. Thus, they are not normally regulated as clinical trials although they are referred to as Phase IV trials in the US.

Definitions

Adverse Reaction/ Adverse Drug Reaction (ADR) (Human drug)
Adverse reaction means a response to a medicinal product which is noxious and unintended and which occurs at doses normally used in man for the prophylaxis, diagnosis or therapy of disease or for the restoration, correction or modification of physiological function.

Serious Adverse Reaction (Human drug)
Serious adverse reaction means an adverse reaction, which results in death, is life-threatening, requires inpatient hospitalisation, or prolongation of existing hospitalisation, results in persistent or significant disability or incapacity, or is a congenital anomaly/birth defect.

Human adverse reaction (Veterinary drug)
A reaction which is noxious and unintended and which occurs in a human being following exposure to a veterinary medicine

Adverse Reaction (Veterinary drug)
A reaction which is harmful and unintended and which occurs at doses normally used in animals for the prophylaxis, diagnosis or treatment of disease or the modification of physiological function.

Serious Adverse Reaction (Veterinary drug)
An adverse reaction which results in death, is life-threatening, results in significant disability or incapacity, is a congenital anomaly/birth defect or which results in permanent or prolonged signs in the animals treated.

Unexpected
Effects not mentioned in the Summary of Product Characteristics or of a more severe nature

Expedited Reporting

Event Location	Human Drugs	Veterinary Drugs
EU	Serious Adverse Reactions	Serious Adverse Reactions Human Adverse Reactions
Non-EU	Serious unexpected Adverse Reactions	Serious unexpected Adverse Reactions Unexpected human Adverse Reactions

Note: The MAH holder is also expected to report serious (unexpected) adverse reactions that occurred when the product was used outside the recommended conditions as defined in the Summary of Product Characteristics e.g. overdose, abuse, etc.
Lack of efficacy in critical situations may also be reported, such as failure of antibiotics to treat life-threatening infections, or vaccine or contraceptive failures.

Figure 12.6 Adverse reaction definitions.

12.6.4
Expedited Reports

The criteria used to identify events that require expedited reporting to European regulatory authorities are shown in Figure 12.6. The maximum time allowed for the submission of an expedited report is 15 days. The clock starts from the time that any of the personnel of the marketing authorisation holder becomes aware of the basic data elements, which consist of:

- an identifiable patient;
- an identifiable reporter;
- a suspect drug; and
- an adverse event.

If the event occurs in the EU, reports should be made to the Competent Authority of the Member State where it occurred, and additionally to the reference member state, if authorised via national procedures. Events occurring outside the EU must be reported to the EMEA and all Member States where the drug is authorised. Reports should be made using harmonised medical terminology as provided in ICH M1MedDRA (Medical Dictionary for Regulatory Activities) guideline, and transmitted preferably using standardised electronic forms, as outlined in the ICH E2B guideline.

12.6.5
Periodic Safety Update Reports

Marketing authorisation holders are obliged to submit Periodic Safety Update Reports (PSURs) according to the frequencies outlined as follows:

- 6-monthly for the first 2 years after authorisation;
- annually for the subsequent 2 years;
- at the time of the first marketing authorisation renewal; and
- thereafter 5-yearly at the time of further renewal.

A PSUR may also be requested by the Competent Authority, should circumstances dictate. The PSURs are intended to provide an update on the global safety experience over the reporting period drawn from all adverse event reports, relevant non-clinical and clinical safety data. The PSURs are expected to contain succinct summary information on all safety issues, together with a critical evaluation of the risk/benefit of the drug in the light of any new or changing post-authorisation information viewed against a background of the usage level. These reports should be submitted to the Competent Authorities of the Member States where the drug is authorised, and additionally to the EMEA in the case of centrally authorised products.

12.6.6
Safety Study Reports

Reports of safety studies must be provided to the regulatory authorities on completion

12.6.7
Response to Issues

National Competent Authorities are primarily responsible for managing responses to pharmacovigilance issues on their territory. However, the reference Member State is expected to take the lead role in evaluating issues for products that have been authorised via national procedures. For community authorisations or drugs that were approved using the EMEA referral procedure, the EMEA will perform this task using the services of the original authorisation rapporteur, a pharmacovigilance working group, and the relevant pharmaceutical committee. In all cases information exchange and communication between the parties is key to achieving a coordinated and measured response. This should include joint actions to ensure that the medical profession, the public or bodies outside the EU are informed of relevant information in a consistent and coherent fashion. Some of the information exchange pathways are shown diagrammatically in Figure 12.7.

For urgent safety issues a Rapid Alert System has been established to ensure that the crisis is managed quickly and efficiently. The EudraNET network is at the core of this response as it provides an efficient means of confidential data exchange between

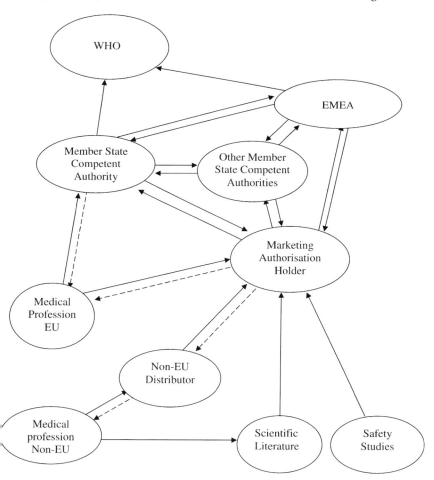

Figure 12.7 Information flows in pharmacovigilance.

the regulatory authorities. The Rapid Alert System should be used in the following circumstances:

- The urgent variation, suspension or withdrawal of the marketing authorisation or recall of the product from the market.

- Significant changes in the Summary of Product Characteristics (SPC) such as, the introduction of new contraindications, the introduction of new warnings, the reduction in the recommended dose, the restriction in the indications, the restriction in the availability of the medicinal product.

- The need to inform healthcare professionals or patients about an identified risk, without delay.

This system is also used where an urgent product recall is required because of safety issues arising from quality problems during manufacture, for example a labelling mix up or the contamination of a sterile product. The authorities must be advised immediately of any situations where such action is contemplated, while a report on the effectiveness of the recall should be furnished at the end of the process

12.6.8
Renewal of Marketing Authorisations

Pharmacovigilance also plays a key role in the renewal of marketing authorisations An initial marketing authorisation is granted for a period of only 5 years; thus, the manufacturing authorisation holder must submit a renewal application 6 months in advance of the expiry of the initial authorisation. This should contain a summary of all relevant information gained from the market since the original authorisation. If reassured that the pharmacovigilance and other data have not identified significant concerns, the authorities will then grant an authorisation for an unlimited period

12.6.9
Reporting Requirements in the US

Similar reporting requirements apply under US regulations, although there are some variations with regards to the frequency and information that must be included in different reports. The FDA are able to take a somewhat more proactive approach in investigating the performance of medicines on the market as they have some funding available to co-sponsor studies of relevance. Their main sources of information are illustrated schematically in Figure 12.8.

12.6.10
Expedited Reports

These include the following:

- 15-day alert report: Reports of serious unexpected adverse experiences must be reported within 15 days of the approval holder or his/her affiliates becoming aware

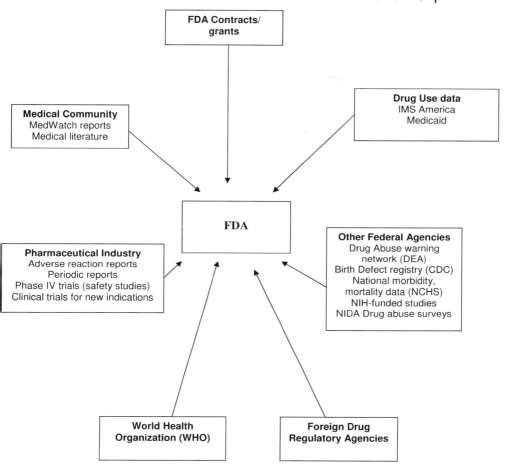

CDC = Center for Disease Control; DEA = Drug Enforcement Agency; NIH = National Institute of Health; NCHS = National Center for Health Statistics; NIDA = National Institute on Drug Abuse; IMS America = commercial source of drug use statistics

Figure 12.8 Drug experience/epidemiological sources available to the FDA (from the CDER Handbook).

of the event. Follow-up reports must be submitted within 15 days of any additional information becoming available.

- Field Alert: Reports of manufacturing defects must be made to the local district office within 3 days of discovery of a problem with a distributed product.

- MedWatch: A centralised MedWatch programme is provided to encourage the voluntary reporting of adverse drug reaction(s) by the medical profession.

Form Approved: OMB No. 0910-0291, Expires: 10/31/08
See OMB statement on reverse.

U.S. Department of Health and Human Services
Food and Drug Administration

For use by user-facilities,
importers, distributors and manufacturers
for MANDATORY reporting

Mfr Report #

UF/Importer Report #

MEDWATCH

FORM FDA 3500A (10/05)

Page _____ of _____

FDA Use Only

A. PATIENT INFORMATION

1. Patient Identifier	2. Age at Time of Event:	3. Sex	4. Weight
	or _____	☐ Female	_____ lbs
In confidence	Date of Birth:	☐ Male	or _____ kgs

B. ADVERSE EVENT OR PRODUCT PROBLEM

1. ☐ Adverse Event   and/or   ☐ Product Problem *(e.g., defects/malfunctions)*

2. Outcomes Attributed to Adverse Event *(Check all that apply)*

☐ Death: _____ *(mm/dd/yyyy)*   ☐ Disability or Permanent Damage

☐ Life-threatening   ☐ Congenital Anomaly/Birth Defect

☐ Hospitalization - initial or prolonged   ☐ Other Serious (Important Medical Events)

☐ Required Intervention to Prevent Permanent Impairment/Damage (Devices)

3. Date of Event *(mm/dd/yyyy)*	4. Date of This Report *(mm/dd/yyyy)*

5. Describe Event or Problem

6. Relevant Tests/Laboratory Data, Including Dates

7. Other Relevant History, Including Preexisting Medical Conditions *(e.g., allergies, race, pregnancy, smoking and alcohol use, hepatic/renal dysfunction, etc.)*

PLEASE TYPE OR USE BLACK INK

C. SUSPECT PRODUCT(S)

1. **Name** *(Give labeled strength & mfr/labeler)*

#1

#2

2. Dose, Frequency & Route Used	3. Therapy Dates *(If unknown, give duration) from/to (or best estimate)*
#1	#1
#2	#2

4. Diagnosis for Use *(Indication)*

#1

#2

6. Lot #	7. Exp. Date
#1	#1
#2	#2

9. NDC# or Unique ID

5. **Event Abated After Use Stopped or Dose Reduced?**
#1 ☐ Yes ☐ No ☐ Doesn't Apply
#2 ☐ Yes ☐ No ☐ Doesn't Apply

8. **Event Reappeared After Reintroduction?**
#1 ☐ Yes ☐ No ☐ Doesn't Apply
#2 ☐ Yes ☐ No ☐ Doesn't Apply

10. Concomitant Medical Products and Therapy Dates *(Exclude treatment of event)*

D. SUSPECT MEDICAL DEVICE

1. Brand Name

2. Common Device Name

3. Manufacturer Name, City and State

4. Model #	Lot #	5. Operator of Device
Catalog #	Expiration Date *(mm/dd/yyyy)*	☐ Health Professional
Serial #	Other #	☐ Lay User/Patient ☐ Other:

6. If Implanted, Give Date *(mm/dd/yyyy)*	7. If Explanted, Give Date *(mm/dd/yyyy)*

8. Is this a Single-use Device that was Reprocessed and Reused on a Patient?
☐ Yes   ☐ No

9. If Yes to Item No. 8, Enter Name and Address of Reprocessor

10. Device Available for Evaluation? *(Do not send to FDA)*
☐ Yes   ☐ No   ☐ Returned to Manufacturer on: _____ *(mm/dd/yyyy)*

11. Concomitant Medical Products and Therapy Dates *(Exclude treatment of event)*

E. INITIAL REPORTER

1. Name and Address	Phone #

2. Health Professional?	3. Occupation	4. Initial Reporter Also Sent Report to FDA
☐ Yes ☐ No		☐ Yes ☐ No ☐ Unk

Submission of a report does not constitute an admission that medical personnel, user facility, importer, distributor, manufacturer or product caused or contributed to the event.

Figure 12.9 FDA form 3500A for mandatory reporting of adverse reactions.

MEDWATCH

FORM FDA 3500A (10/05) *(continued)*   Page ____ of ____

FDA USE ONLY

F. FOR USE BY USER FACILITY/IMPORTER *(Devices Only)*

1. Check One		2. UF/Importer Report Number
☐ User Facility	☐ Importer	

3. User Facility or Importer Name/Address

4. Contact Person	5. Phone Number

6. Date User Facility or Importer Became Aware of Event *(mm/dd/yyyy)*	7. Type of Report	8. Date of This Report *(mm/dd/yyyy)*
	☐ Initial	
	☐ Follow-up # ____	

9. Approximate Age of Device	10. Event Problem Codes *(Refer to coding manual)*		
	Patient Code	☐	☐
	Device Code	☐	☐

11. Report Sent to FDA?
☐ Yes _____ *(mm/dd/yyyy)*
☐ No

12. Location Where Event Occurred
☐ Hospital
☐ Home
☐ Nursing Home
☐ Outpatient Treatment Facility
☐ Outpatient Diagnostic Facility
☐ Ambulatory Surgical Facility
☐ Other: _____ *(Specify)*

13. Report Sent to Manufacturer?
☐ Yes _____ *(mm/dd/yyyy)*
☐ No

14. Manufacturer Name/Address

G. ALL MANUFACTURERS

1. Contact Office - Name/Address *(and Manufacturing Site for Devices)*	2. Phone Number

3. Report Source *(Check all that apply)*
☐ Foreign
☐ Study
☐ Literature
☐ Consumer
☐ Health Professional
☐ User Facility
☐ Company Representative
☐ Distributor
☐ Other: _____

4. Date Received by Manufacturer *(mm/dd/yyyy)*	5.
	(A)NDA # _____
6. If IND, Give Protocol #	IND # _____
	STN # _____
	PMA/ 510(k) # _____

7. Type of Report *(Check all that apply)*
☐ 5-day ☐ 30-day
☐ 7-day ☐ Periodic
☐ 10-day ☐ Initial
☐ 15-day ☐ Follow-up # ____

Combination Product ☐ Yes _____
Pre-1938 ☐ Yes _____
OTC Product ☐ Yes _____

9. Manufacturer Report Number	8. Adverse Event Term(s)

H. DEVICE MANUFACTURERS ONLY

1. Type of Reportable Event
☐ Death
☐ Serious Injury
☐ Malfunction
☐ Other:

2. If Follow-up, What Type?
☐ Correction
☐ Additional Information
☐ Response to FDA Request
☐ Device Evaluation

3. Device Evaluated by Manufacturer?
☐ Not Returned to Manufacturer
☐ Yes ☐ Evaluation Summary Attached
☐ No *(Attach page to explain why not)* or provide code:

4. Device Manufacture Date *(mm/yyyy)*

5. Labeled for Single Use?
☐ Yes ☐ No

6. Evaluation Codes *(Refer to coding manual)*			
Method	☐ –	☐ –	☐ –
Results	☐ –	☐ –	☐ –
Conclusions	☐ –	☐ –	☐ –

7. If Remedial Action Initiated, Check Type
☐ Recall
☐ Repair
☐ Replace
☐ Relabeling
☐ Other: _____
☐ Notification
☐ Inspection
☐ Patient Monitoring
☐ Modification/ Adjustment

8. Usage of Device
☐ Initial Use of Device
☐ Reuse
☐ Unknown

9. If action reported to FDA under 21 USC 360i(f), list correction/ removal reporting number:

10. ☐ Additional Manufacturer Narrative   and / or   11. ☐ Corrected Data

---

Department of Health and Human Services
Food and Drug Administration - MedWatch
10903 New Hampshire Avenue
Building 22, Mail Stop 4447
Silver Spring, MD 20993-0002
Please DO NOT RETURN this form to this address.

Figure 12.9 *(Continued)*

Table 12.2 Forms to be use for expedited reporting of adverse events.

Form No	Purpose
Form FDA 3500	Voluntary reporting of adverse events (reactions, failures, malfunctions) with drugs or devices by medical personnel
Form FDA 3500A	Mandatory reporting of adverse drug events by manufacturers
	Mandatory reporting of adverse device events by manufacturers, importers and device user facilities
Form VAERS-1	Mandatory reporting of Vaccine Adverse Events by manufacturers
Form FDA 1932	Mandatory reporting of Veterinary Adverse Drug Reactions, Lack of Effectiveness, or Product Defects by manufacturers

The forms used for expedited reporting of adverse events are listed in Table 12.2 An example of mandatory reporting form FDA3500A is shown in Figure 12.9.

12.6.11
Periodic Reports

Depending on the type of product, differences in the frequency and content of information in periodic reports apply; this is summarised in Table 12.3.

12.7
Advertising and Promotion

The regulations also address the need to ensure that drug information provided by pharmaceutical firms is truthful, balanced, and accurately communicated. As such, i must be consistent with the indications for use and the established performance and limitations. In Europe, the directives do not impose a specific requirement to review advertising or promotional material before it is released. Acceptable standards may be achieved via voluntary codes of practice and self-regulation. However, nationa authorities must monitor such material and should have the power to act where the need arises. In the US, the FDA must vet advertising and promotional material before it is released.

12.8
Market Vigilance and Oversight of Devices

Just as with drugs, market vigilance systems are required to ensure that problem with devices are identified and addressed in a timely manner. Although they share

Table 12.3 Summary of periodic reporting obligations in the US.

Report type	Frequency	Information included
Human drugs		
Periodic adverse experience report 21 CFR 314.80 (c) 2 21 CFR 600.80 (c) 2	Every 3 months for the first 3 years; annually thereafter	Summary and analysis of all adverse incidents Reports of non-15-day adverse incidents History of actions taken in response to incidents
Distribution report – Biologics only 21 CFR 600.81	Every 6 months	Distribution data
Annual Report 21 CFR 314.81 (b) 2 21 CFR 601.28	Annually	Summary of significant information from previous report Distribution data Current labelling CMC new information New non-clinical studies New clinical data Status report on post-approval study commitments, e.g. paediatric studies Minor supplementary changes
Veterinary drugs		
Periodic Drug Experience Report 21 CFR 514.80 (b) 4	Every 6 months for the first 2 years; annually thereafter	Distribution data Current Labelling New non-clinical and clinical data Adverse events not previously reported Summary of increased adverse event frequency Minor supplementary changes

many of the essential features of pharmacovigilance systems in terms of communication channels, there are obvious differences in terms of reporting criteria and requirements.

12.8.1
Market Vigilance in the EU

The responsibilities of Competent Authorities in terms of market vigilance are set out in the Articles of the device directives, whereas the obligations for manufacturers are contained in the Annexes. MEDDEV guide 2.12-1 provides detailed practical guidance on how the medical device vigilance system should operate within the EU, the European Economic Area (EEA) and Switzerland. This guidance was significantly updated in 2007 to take account of guidance put forward by the Global Harmonisa-tion Task Force (GHTF) on international vigilance and post-market surveillance systems, and to address the introduction of the EUDAMED database for medical devices. The medical profession and other users of devices are the primary source of information relating to problems with devices on the market. It is expected that the manufacturer will receive information on such issues through complaints proce-dures. Additionally, some Member States may also require the medical profession to submit reports directly to the Competent Authority. The manufacturer must then evaluate the information against the following criteria to see if there is an obligation to report the incident to the Competent Authority:

- an event has occurred;
- the manufacturer's device is suspected to be a contributory cause of the incident;
- the incident led, or might have led, to either the death or serious deterioration in the health of a patient, user or other person.

A serious deterioration in health can include:

- life-threatening illness;
- permanent impairment of a body function or permanent damage to a body structure;
- a condition necessitating medical or surgical intervention to prevent life-threaten-ing illness, permanent impairment of a body function, or permanent damage to a body structure;
- any indirect harm as a consequence of an incorrect diagnostic or IVD test results when used within manufacturer's instructions for use; and
- foetal distress, foetal death or any congenital abnormality or birth defects.

Manufacturers should also report incidents where user error resulted in death or a serious deterioration in the state of health, or created a serious threat to public health. The manufacturer is also obliged to monitor trends and report where a significant increase in the level of incidents is observed, even if individual incidents would not be reportable in isolation. Abnormal use events should be addressed to the healthcare facility where they occur. The manufacturer should endeavour to report incidents immediately, once they become aware of the suspected involvement of their device, but in any event the following time limits must be respected:

- Serious public health threat: 2 days
- Death or unanticipated serious deterioration in state of health: 10 days
- Others: 30 days.

Incidents should be reported to the Competent Authority of the country where the occurrence took place, using the form illustrated in Figure 12.10.

The manufacturer will then conduct a full investigation of the incident to determine the cause and the actions that must be taken in order to rectify the situation and/or prevent a reoccurrence. This may often result in a Field Safety Corrective Action (FSCA), whereby devices should be removed or modified in the field to avoid the risk of death or serious deterioration in the state of health associated

V.05/07

Report Form
Manufacturer's Incident Report

Medical Devices Vigilance System
(MEDDEV 2.12/1 rev 5)

1. Administrative information
Recipient
Name of national competent authority (NCA)
Address of national competent authority
Date of this report
Reference number assigned by the manufacturer
Reference number assigned by NCA to whom sent (if known)

Type of report
- ☐ Initial report
- ☐ Follow-up report
- ☐ Combined initial and final report
- ☐ Final report

Classification of incident
- ☐ Death or unanticipated serious deterioration in state of health, serious public health threat
- ☐ All other reportable incidents

Identify to what other NCAs this report was also sent

2 Information on submitter of the report

Status of submitter
- ☐ Manufacturer
- ☐ Authorised representative within EEA and Switzerland
- ☐ Others (identify the role):

3 Manufacturer information

Manufacturer name

Manufacturer's contact person

Address

Postal code	City
Phone	Fax
E-mail	Country

4 Authorised Representative information

Name of the authorised representative

The authorised representative's contact person

Address

Postal code	City

Figure 12.10 Manufacturer's Incident Report Form.

Phone	Fax
E-mail	Country

5 Submitter's information (if different from section 3 or 4)

Submitter's name

Name of the contact person

Address

Postal code	City
Phone	Fax
E-mail	Country

6 Medical device information

Class

☐   AIMD Active implants

☐ MDD Class III	☐ IVD Annex II List A
☐ MDD Class IIb	☐ IVD Annex II List B
☐ MDD Class IIa	☐ IVD Devices for self-testing
☐ MDD Class I	☐ IVD General

Nomenclature system (preferable GMDN)

Nomenclature code

Nomenclature text

Commercial name/brand name/make

Model and/or catalogue number

Serial number(s) and/or lot/batch number(s)

Software version number (if applicable)

Manufacturing date/expiry date (if applicable)

Accessories/associated device (if applicable)

Notified body (NB) ID- number

7 Incident information

User facility report reference number, if applicable

Manufacturers awareness date

Date of incident occurred

Incident description narrative

Figure 12.10 (*Continued*)

Number of patients involved (if known)	Number of medical devices involved (if known)
Medical device current location/disposition (if known)	

Operator of the medical device at the time of incident (select one)
☐ health care professional
☐ patient
☐ other

Usage of the medical device (select from list below)	
☐ initial use	☐ reuse of a single use medical device
☐ reuse of a reusable medical device	☐ re-serviced/refurbished
☐ other (please specify):	☐ problem noted prior use

8 Patient information

Patient outcome
Remedial action taken by the healthcare facility relevant to the care of the patient
Age of the patient at the time of incident, if applicable

Gender, if applicable
☐   Female   ☐   Male
Weight in kilograms, if applicable

9 Healthcare facility information

Name of the healthcare facility	
Contact person within the facility	
Address	
Postal code	City
Phone	Fax
E-mail	Country

10 Manufacturer's preliminary comments (Initial/Follow-up report)

Manufacturer's preliminary analysis
Initial corrective actions/preventive actions implemented by the manufacturer
Expected date of next report

11 Results of manufacturers final investigation (Final report)

The manufacturer's device analysis results
Remedial action/corrective action/preventive action/Field Safety Corrective Action
NOTE: In the case of a FSCA the submitter needs to fill in the form of Annex 4

Figure 12.10 (*Continued*)

Time schedule for the implementation of the identified action

Final comments from the manufacturer

Further investigations

Is the manufacturer aware of similar incidents with this type of medical device with a similar root cause?

☐ Yes ☐ No

If yes, state in which countries and the report reference numbers of the incidents

For final report only. The medical device has been distributed to the following countries:

Within EEA and Switzerland:

☐ AT ☐ BE ☐ BU ☐ CH ☐ CY ☐ CZ ☐ DE ☐ DK ☐ EE ☐ ES
☐ FI ☐ FR ☐ GB ☐ GR ☐ HU ☐ IE ☐ IS ☐ IT ☐ LI ☐ LT
☐ LU ☐ LV ☐ MT ☐ NL ☐ NO ☐ PL ☐ PT ☐ RO ☐ SE ☐ SI
☐ SK

Candidate Countries:

☐ CR ☐ TR

☐ All EEA, Candidate Countries and Switzerland

Others:

12 Comments

I affirm that the information given above is correct to the best of my knowledge.

....
Signature

Name   City   Date

Submission of this report does not, in itself, represent a conclusion by the manufacturer and/or authorized representative or the national competent authority that the content of this report is complete or accurate, that the medical device(s) listed failed in any manner and/or that the medical device(s) caused or contributed to the alleged death or deterioration in the state of the health of any person.

Figure 12.10 (*Continued*)

with their use. The manufacturer must simultaneously inform the Competent Authorities of all the countries affected of the intended FSCA, using the form shown in Figure 12.11. Notifications are also required where a FSCA ensues from incidents occurring in countries that are not participants in the European vigilance system. The notification should include an explanation of the background and justification for the action together with a copy of the actual Field Safety Notice (FSA) that will be issued to those required to take action in the field. A template for a FSA has been developed and is shown in Figure 12.12.

V.05/07

Report Form
Field Safety Corrective Action

Medical Devices Vigilance System
(MEDDEV 2.12/1 rev 5)

1. Administrative information   Destination
Name of national competent authority (NCA)
Address of national competent authority
Date of this report
Reference number assigned by the manufacturer
Incidence reference number and name of the co-ordinating national competent authority (if applicable)
Identify to what other national competent authorities this report was also sent

2 Information on submitter of the report
Status of submitter
☐   Manufacturer
☐   Authorised representative within EEA
☐   Others (identify the role):

3 Manufacturer information	
Manufacturer name	
Manufacturer's contact person	
Address	
Postal code	City
Phone	Fax
E-mail	Country

4 Authorised representative information	
Name of the authorised representative	
The authorised representative's contact person	
Address	
Postal code	City
Phone	Fax
E-mail	Country

Figure 12.11 FSCA Report Form.

5 National contact point information	
National contact point name	
Name of the contact person	
Address	
Postal code	City
Phone	Fax
E-mail	Country

6 Medical device information	
Class ☐ AIMD Active implants	
☐ MDD Class III	☐ IVD Annex II List A
☐ MDD Class IIb	☐ IVD Annex II List B
☐ MDD Class IIa	☐ IVD Devices for self-testing
☐ MDD Class I	☐ IVD General
Nomenclature system (preferable GMDN)	
Nomenclature code	
Nomenclature text	
Commercial name/brand name/make	
Model number	
Serial number(s) and/or lot/batch number(s)	
Software version number (if applicable)	
Manufacturing date/expiry date (if applicable)	
Accessories/associated device (if applicable)	
Notified body (NB) ID- number	

7 Description of FSCA
Background information and reason for the FSCA
Description and justification of the action (corrective/preventive)
Advice on actions to be taken by the distributor and the user
Attached please find ☐ Field Safety Notice (FSN) in English ☐ FSN in national language ☐ Others (please specify):

Figure 12.11 (*Continued*)

Time schedule for the implementation of the different actions

These countries within the EEA and Switzerland are affected by this FSCA

Within EEA and Switzerland:

☐ AT	☐ BE	☐ BU	☐ CH	☐ CY	☐ CZ	☐ DE	☐ DK	☐ EE	☐ ES
☐ FI	☐ FR	☐ GB	☐ GR	☐ HU	☐ IE	☐ IS	☐ IT	☐ LI	☐ LT
☐ LU	☐ LV	☐ MT	☐ NL	☐ NO	☐ PL	☐ PT	☐ RO	☐ SE	☐SI
☐SK									

Candidate Countries:

☐ CR   ☐ TR

☐ All EEA, Candidate Countries and Switzerland

Others:

These countries outside the EEA and Switzerland are affected by this FSCA

8 Comments

I affirm that the information given above is correct to the best of my knowledge.

....
Signature

Name   City   Date

Submission of this report does not, in itself, represent a conclusion by the manufacturer and/or authorized representative or the national competent authority that the content of this report is complete or accurate, that the medical device(s) listed failed in any manner and/or that the medical device(s) caused or contributed to the alleged death or deterioration in the state of the health of any person.

Figure 12.11 (*Continued*)

The Competent Authority to which the incident was originally reported usually takes on the role of monitoring and evaluating how the incident is dealt with. When all actions are complete, including any necessary Field Safety Correction Actions, the manufacturer must submit a final report to the Competent Authority. Under a safeguard clause, Competent Authorities may take unilateral action to remove a device from their market, if they believe that it poses an unacceptable health risk, but the Commission must be informed of such actions. In less-urgent situations, concerns may be referred to the Committee for Medical Devices for consideration.

12.8.2
Medical Device Vigilance in the US

The requirements for market vigilance and oversight in the US are set out in 21 CFR 803, Medical Device Reporting, 21 CFR 806, Reports of Corrections and Removals and 21 CFR 822 Post market Surveillance.

Urgent Field Safety Notice

Commercial name of the affected product
FSCA-identifier (e.g. date)
Type of action (e.g. Refer to definition of a FSCA in MEDDEV 2.12/1)

Date:
Attention: ///////////////

Details on affected devices:
Specific details to enable the affected product to be easily identified: e.g. type of device,
model name and number, batch/ serial numbers of affected devices and part or order
number.
Insert or attach list of individual devices.
(Possible reference to a manufacturer web site.)

Description of the problem:
A factual statement explaining the reasons for the FSCA, including description of the device
deficiency or malfunction, clarification of the potential hazard associated with the continued
use of the device and the associated risk to the patient, user or other person.
Any possible risk to patients associated with previous use of affected devices.

Advise on action to be taken by the user:
Include, as appropriate:
• *identifying and quarantining the device*
• *method of recovery, disposal or modification of device*
• *recommended patient follow up, e.g. implants, IVD*
• *timelines*
• *Confirmation form to be sent back to the manufacturer if an action is required (e.g. return of*
  products)

Transmission of this Field Safety Notice: (if appropriate)
This notice needs to be passed on all those who need to be aware within your organization or to
any organisation where the potentially affected devices have been transferred. (If appropriate)
Please transfer this notice to other organisations on which this action has an impact (if
appropriate).
Please maintain awareness on this notice and resulting action for an appropriate period to ensure
effectiveness of the corrective action (if appropriate).

Contact reference person:
Name/organisation, address, contact details.

The undersign confirms that this notice has been notified the appropriate Regulatory Agency
(Closing paragraph)
Signature

Figure 12.12 Field Safety Action Template.

12.8.3
Medical Device Reporting

Medical device manufacturers, importers and user facilities are obliged to report
adverse incidents as summarised in Table 12.4. The criteria for what constitutes a

Table 12.4 Summary of Medical Device Reporting (MDR) requirements for individual adverse incidents.

Reporter	Time allowed (days)	Report to FDA	Report to Manufacturer
User Facility	10	Death	Death Serious injury
Importer	30	Death Serious injury	Death Serious injury Malfunctions
Manufacturer	30	Death Serious injury Malfunctions	–
Manufacturer	5	Events requiring remedial action to prevent public health threat	–

reportable event are essentially the same as in Europe. Reports must be made on MedWatch form 5300A, which is the same form as used for mandatory reporting of adverse drug reactions. Manufacturers are required to report events within 5 days in circumstances where remedial action is required to prevent an unreasonable risk of significant harm to public health, or where the FDA make a specific request due to concern over a possible problem. User facilities and manufacturers are also obliged to submit annual reports/baseline reports to the FDA. Records of medical device reporting events, including the evaluation and decision-making process, must be maintained for a minimum of 2 years, or for the expected life of the device in the case of manufacturers where the files may be integrated into the complaint files. Device distributors must also maintain records of events as part of their complaint files, but are not required to submit reports to the FDA.

12.8.4
Reports of Corrections and Removals

Corrections or removals of devices from the market may be necessitated either as a consequence of adverse event reports or the discovery of manufacturing or other defects that pose a risk to public health. The manufacturer must submit a report to the FDA within 10 days of initiating such corrections or removals. This should provide information on the identity and number of devices concerned, the reasons for doing so, and the communication of the action. The manufacturer must also maintain records of other corrections or removals that need not be reported.

12.8.5
Post-Market Surveillance

The FDA may also require manufacturers to conduct post-market surveillance studies for Class II and III devices that meet the following criteria:

- Failure of the device would be reasonably likely to have serious adverse health consequences.
- The device is intended to be implanted in the human body for more than 1 year.
- The device is intended to be used outside a user facility to support or sustain life.

Surveillance requests are usually targeted at particular concerns, which the FDA will specify as part of the scope of the surveillance order. The manufacturer must submit a plan to the FDA outlining how the surveillance will be conducted. Once approved, the surveillance plan must be put into effect, usually for not more than 3 years, and any reports agreed as part of the plan must be submitted to the FDA. Surveillance plans are quite similar in concept to the safety studies that may be agreed with regulatory authorities for drugs.

12.9
Chapter Review

In this chapter, attention has been focussed on the regulatory mechanisms in place to ensure that consistently high-quality product is manufactured on a routine basis, and that any issues which emerge on the market are identified and dealt with in a timely manner. The registration of manufacturers and regular site inspections are key to assuring that appropriate manufacturing standards are maintained. Early detection of problems on the market is dependent on the effective operation of vigilance systems and the cooperation of medical, industrial and regulatory personnel. Although differences in the detailed practices exist, the same basic regulatory concepts are applied in Europe and the US.

12.10
Further Reading

- European Regulations
  Veterinary Medicines Directive 2001/82/EC
  o Title IV: Manufacture and imports
  o Title VII: Pharmacovigilance
  o Title VIII: Supervision and sanctions.

  Human Medicines Directive 2001/83/EC
  o Title IV: Manufacture and imports,
  o Title VIII: Advertising, Title IX: Pharmacovigilance,
  o Title XI: Supervision and sanctions

  Medicinal Products Regulation (EC) No. 726/2004
  o Title II, Chapter 2 Supervision and penalties; Chapter 3, Pharmacovigilance
  o Title III, Chapter 2 Supervision and penalties; Chapter 3, Pharmacovigilance

  http://ec.europa.eu/enterprise/pharmaceuticals/eudralex/index.htm.

Active Implantable Medical Devices Directive 90/385 EEC Article 8 (Competent Authority Vigilance procedure), Annex II 3.1 (manufacturer's vigilance and reporting duties)

Medical Devices Directive 93/42 EEC, Article 10 (Competent Authority Vigilance procedure), Article 14 (Manufacturer registration), Annex II 3.1 (i) & (ii) manufacturer's vigilance and reporting duties)

In Vitro Diagnostics Directive 98/79/EC, Article 10 (Registration of manufacturers), Article 11 (Vigilance procedure Competent Authorities), Annex III 5 (manufacturer's vigilance and reporting duties)

http://ec.europa.eu/enterprise/medical_devices/index_en.htm.

- US Regulations
  FDC Act Section 510, Registration of producers of drugs and devices
  Establishment Registration: 21 CFR Parts 207 (drugs), 607 (blood/blood products) and 807 (devices)
  Drug Reporting requirements: 21 CFR Parts 314.80, 314.81, 514.80, 600.80, 600.81, 601.28
  Medical Devices: 21 CFR 803, Medical Device Reporting, 21 CFR 806, Reports of Corrections and Removals and 21 CFR 822 Post market Surveillance
  www.fda.gov.

- Guidance Documents
  The Rules Governing Medicinal Products in the European Union, Volume 9, Pharmacovigilance
  http://ec.europa.eu/enterprise/pharmaceuticals/eudralex/index.htm.
  MEDDEV guide 2.12-1 Guidelines on a Medical Devices Vigilance System
  http://ec.europa.eu/enterprise/medical_devices/meddev/index.htm.
  EMEA Compilation of Community Procedures on Inspections and Exchange of Information – Conduct of Inspections of Pharmaceutical Manufacturers
  http://www.emea.europa.eu/htms/aboutus/emeaoverview.htm.
  FDA Compliance Programme Guidance Manual Programme 7536.002, Drug Manufacturing Inspections
  FDA Guide to Inspections of Quality Systems (QSIT Guide)
  CDER Handbook
  www.fda.gov.

Index

Medical Product Regulatory Affairs. John J. Tobin and Gary Walsh
Copyright © 2008 WILEY-VCH Verlag GmbH & Co. KGaA, Weinheim
ISBN: 978-3-527-31877-3

Printed and bound by CPI Group (UK) Ltd, Croydon, CR0 4YY